T0182129

Irregularities in the Distribution of Prime Numbers

János Pintz • Michael Th. Rassias
Editors

Irregularities in the Distribution of Prime Numbers

From the Era of Helmut Maier's Matrix Method and Beyond

 Springer

Editors
János Pintz
Alfréd Rényi Institute of Mathematics
Hungarian Academy of Sciences
Budapest, Hungary

Michael Th. Rassias
Institute of Mathematics
University of Zürich
Zürich, Switzerland

Moscow Institute of Physics
and Technology
Dolgoprudny, Russia

Institute for Advanced Study
Program in Interdisciplinary Studies
Princeton, NJ, USA

ISBN 978-3-030-06514-0   ISBN 978-3-319-92777-0   (eBook)
https://doi.org/10.1007/978-3-319-92777-0

Mathematics Subject Classification (2010): 11-XX, 05-XX

Printed on acid-free paper

This Springer imprint is published by the registered company Springer International Publishing AG part of Springer Nature.
The registered company address is: Gewerbestrasse 11, 6330 Cham, Switzerland

Foreword

Although prime numbers have been of interest to mathematicians since ancient times, reasonably precise rules on their distribution were found relatively late. The first known successful guess was accomplished by Gauss who announced in 1849 that he had come to the conclusion that "at around x, the primes occur with density $1/\log x$" [10]. He concluded that $\pi(x)$ could be approximated by

$$Li(x) := \int_2^x \frac{dt}{\log t} = \frac{x}{\log x} + \frac{x}{\log^2 x} + O\left(\frac{x}{\log^3 x}\right).$$

A sharp version of this result, the Prime Number Theorem with error term, could be proven by Hadamard [11] and de le Vallée-Poussin [2] in 1896. Cramér in 1936 interpreted Gauss' statement in terms of probability theory: He considered a sequence T of independent random variables $Z_2, Z_3, \ldots$ with

$$\mathrm{Prob}(Z_n = 1) = \frac{1}{\log n} \quad \text{and} \quad \mathrm{Prob}(Z_n = 0) = 1 - \frac{1}{\log n}$$

and suggested that many relations true for the random sequence T with probability 1 should also hold true for the sequence of prime numbers. This principle suggests that

$$\pi(x + y) - \pi(x) \sim \frac{y}{\log x}, \quad \text{provided} \quad \frac{y}{\log^2 x} \to \infty \text{ as } x \to \infty.$$

It was proved by Selberg [19] that this indeed holds true for almost all x, if one assumes the Riemann Hypothesis. In 1985 the author [14] could show that here Cramér's model leads to false predictions. For the proof he applied the matrix method, which he had developed in 1981 [13] to establish the existence of chains of large gaps between consecutive primes. The matrix $\mathcal{M}$ is constructed as follows:

One starts with a base row, a subinterval $[w, w + U]$ of the interval $(0, P(x))$, $P(x)$ a primorial, with a suitable number of integers coprime to $P(x)$. The rows of the matrix $\mathcal{M}$ are the translates of this base row:

$$\mathcal{M} := (r P(x) + u)_{\substack{w \leq u \leq w+U \\ R \leq r < 2R}} .$$

Then the number of primes in $\mathcal{M}$ is counted by applying results on primes in arithmetic progressions to the columns of $\mathcal{M}$. At least one row of the matrix $\mathcal{M}$ then contains the desired configuration of prime numbers. Another ingredient of the above proof was results from the theory of differential-difference equations whose study had been initiated by N. G. de Bruijn [1].

In the sequel the matrix method has been applied by several authors to obtain other irregularity results: irregularity results for the distribution of prime numbers in arithmetic progressions and in polynomial sequences. The matrix method has been combined with analytic methods, e.g., the saddle point method (see [5–8, 12, 17]). There also have been applications to other rings of algebraic objects [20]. Some irregularity results have been discovered by different methods [18]. The matrix method has been applied to make progress towards the twin prime conjecture [15]. Recently several authors have combined the matrix method with methods initiated by Goldston, Pintz, and Yildirim [9] and continued by Ford, Green, Konyagin, Maynard, and Tao ([3], see also Maier and Rassias [16]), as well as Ford, Maynard, and Tao [4].

Many colleagues who have been involved in the research just described now have contributed to the present volume. The author wishes to express his deep gratitude to them and also to the other contributors of this volume. My special thanks go to János Pintz and Michael Th. Rassias who have prepared this book. I also wish to express my thanks to M. Th. Rassias for a long and successful collaboration.

References

1. N.G. de Bruijn, On the number of uncancelled elements in the sieve of Eratosthenes. Nederl. Akad. Wetensch. Proc. **53**, 803–812 (1950)
2. Ch. de la Vallée-Poussin, Recherches analytiques sur la théorie des nombres premiers. Ann. Soc. Sci. Brux. **20**(2), 183–256, 281–297 (1896)
3. K. Ford, B.J. Green, S. Konyagin, J. Maynard, T. Tao, Large gaps between primes. J. Am. Math. Soc. **31**(1), 65–105 (2018)
4. K. Ford, J. Maynard, T. Tao, Chains of large gaps between primes, in *Irregularities in the Distribution of Prime Numbers: From the Era of Helmut Maier's Matrix Method and Beyond* (Springer, Cham, 2018)
5. J. Friedlander, A. Granville, Limitations to the equidistribution of primes I. Ann. Math. **129**, 363–382 (1989)
6. J. Friedlander, A. Granville, Limitations to the equidistribution of primes IV. Proc. R. Soc. Lond. Ser. A **435**, 197–204 (1991)
7. J. Friedlander, A. Granville, Limitations to the equidistribution of primes III. Compos. Math. **81**, 19–32 (1992)

8. J. Friedlander, A. Granville, A. Hildebrand, H. Maier, Oscillation theorems for primes in arithmetic progressions and for sifting functions. J. AMS **4**(1), 25–86 (1991)
9. D.A. Goldston, J. Pintz, C. Yildirim, Primes in tuples I. Ann. Math. **170**, 819–862 (2009)
10. A. Granville, Unexpected irregularities in the distribution of prime numbers, in *Proceedings of International Congress of Mathematics*, Zürich (1994), pp. 388–399
11. J. Hadamard, Sur la distribution des zéros de la fonction $\zeta(s)$ et ses conséquences arithmétiques. Bull. Soc. Math. France **24**, 199–220 (1896)
12. A. Hildebrand, H. Maier, Irregularities in the distribution of primes in short intervals. J. Reine Angew. Math. **397**, 162–193 (1989)
13. H. Maier, Chains of large gaps between consecutive primes. Adv. Math. **39**, 257–269 (1981)
14. H. Maier, Primes in short intervals. Mich. Math. J. **32**(2), 221–225 (1985)
15. H. Maier, Small differences between prime numbers. Mich. Math. J. **35**(3), 323–344 (1988)
16. H. Maier, M.Th. Rassias, Large gaps between consecutive prime numbers containing perfect k-th powers of prime numbers. J. Funct. Anal. **272**, 2659–2696 (2017)
17. A. Nair, A. Perelli, On the prime ideal theorem and irregularities in the distribution of primes. Duke Math. J. **77**, 1–20 (1995)
18. J. Pintz, Cramér vs. Cramér. On Cramér's probabilistic model for primes. Funct. Approx. Comment. Math. **37**(2), 361–376 (2007)
19. A. Selberg, On the normal density of primes in small intervals, and the difference between consecutive primes. Arch. Math. Naturvid. **47**(6), 87–105 (1943)
20. F. Thorne, Maier matrices beyond $\mathbb{Z}$, in *Combinatorial Number Theory (Proceedings of the Integers Conference)* (Walter de Gruyter, Berlin, 2007), pp. 185–192

Department of Mathematics   Helmut Maier
University of Ulm
Ulm, Germany

March 18, 2018

Preface

Irregularities in the Distribution of Prime Numbers: From the Era of Maier's Matrix Method and Beyond presents research as well as expository papers under the broad banner of the vibrant and fascinating domain of the study of irregularities in the distribution of primes. The papers featured in this book have been contributed by experts from the international community. A great deal of research in this area has been inspired and channeled by the now famous Maier matrix method, which was introduced in the 1980s in a sensational paper of Helmut Maier. Maier's pioneering work in the domain established a newfound understanding about the distribution of primes, radically contradicting what was expected by probabilistic models until that time.

The editors would like to express their gratitude to all the contributors of book chapters, who participated in this collective effort.

We would also like to express our thanks to the staff of Springer for our excellent collaboration for the production of this book.

Budapest, Hungary   János Pintz
Zürich, Switzerland   Michael Th. Rassias

Contents

Chains of Large Gaps Between Primes

Kevin Ford, James Maynard, and Terence Tao

Abstract Let p_n denote the n-th prime, and for any $k \geqslant 1$ and sufficiently large X, define the quantity

$$G_k(X) := \max_{p_{n+k} \leqslant X} \min(p_{n+1} - p_n, \ldots, p_{n+k} - p_{n+k-1}),$$

which measures the occurrence of chains of k consecutive large gaps of primes. Recently, with Green and Konyagin, the authors showed that

$$G_1(X) \gg \frac{\log X \log\log X \log\log\log\log X}{\log\log\log X}$$

for sufficiently large X. In this note, we combine the arguments in that paper with the Maier matrix method to show that

$$G_k(X) \gg \frac{1}{k^2} \frac{\log X \log\log X \log\log\log\log X}{\log\log\log X}$$

for any fixed k and sufficiently large X. The implied constant is effective and independent of k.

K. Ford (✉)
Department of Mathematics, University of Illinois at Urbana-Champaign, Urbana, IL, USA
e-mail: ford@math.uiuc.edu

J. Maynard
Mathematical Institute, Radcliffe Observatory Quarter, Oxford, England

T. Tao
Department of Mathematics, UCLA, Los Angeles, CA, USA
e-mail: tao@math.ucla.edu

© Springer International Publishing AG, part of Springer Nature 2018
J. Pintz, M. Th. Rassias (eds.), *Irregularities in the Distribution of Prime Numbers*,
https://doi.org/10.1007/978-3-319-92777-0_1

1 Introduction

Let p_n denote the nth prime, and for any $k \geqslant 1$ and sufficiently large X, let

$$G_k(X) := \max_{p_{n+k} \leqslant X} \min(p_{n+1} - p_n, \ldots, p_{n+k} - p_{n+k-1}),$$

denote the maximum gap between k consecutive primes less than X. The quantity $G_1(X)$ has been extensively studied. The prime number theorem implies that

$$G_1(X) \geqslant (1 + o(1)) \log X,$$

with the bound being successively improved in many papers [1, 4, 9–11, 15, 17, 19–23]. The best lower bound currently is[1]

$$G_1(X) \gg \frac{\log X \log_2 X \log_4 X}{\log_3 X},$$

for sufficiently large X and an effective implied constant, due to [11]. This result may be compared against the conjecture $G_1(X) \asymp \log^2 X$ of Cramér [6] (see also [13]), or the upper bound $G_1(X) \ll X^{0.525}$ of Baker-Harman-Pintz [3], which can be improved to $G_1(X) \ll X^{1/2} \log X$ on the Riemann hypothesis [5].

Now we turn to $G_k(X)$ in the regime where $k \geqslant 1$ is fixed, and X assumed sufficiently large depending on k. Clearly $G_k(X) \leqslant G_1(X)$, and a naive extension of the probabilistic heuristics of Cramér [6] suggests that $G_k(X) \asymp \frac{1}{k} \log^2 X$ as $X \to \infty$. The first non-trivial bound on $G_k(X)$ for $k \geqslant 2$ was by Erdős [9], who showed that

$$G_2(X)/\log X \to \infty$$

as $X \to \infty$. Using what is now known as the Maier matrix method, together with the arguments of Rankin [20] on $G_1(X)$, Maier [14] showed that

$$G_k(X) \gg_k \frac{\log X \log_2 X \log_4 X}{(\log_3 X)^2}$$

for any fixed $k \geqslant 1$ and a sequence of X going to infinity. Recently, by modifying Maier's arguments and using the more recent work on $G_1(X)$ in [10, 17], this was improved by Pintz [18] to show that

$$G_k(X)/\left(\frac{\log X \log_2 X \log_4 X}{(\log_3 X)^2}\right) \to \infty$$

for a sequence of X going to infinity.

---

[1] As usual in the subject, $\log_2 x := \log\log x$, $\log_3 x := \log\log\log x$, and so on. The conventions for asymptotic notation such as $\ll$ and $o()$ will be defined in Sect. 1.1.

Our main result here is as follows.

Theorem 1 *Let $k \geqslant 1$ be fixed. Then for sufficiently large X, we have*

$$G_k(X) \gg \frac{1}{k^2} \frac{\log X \log_2 X \log_4 X}{\log_3 X}.$$

The implied constant is absolute and effective.

Maier's original argument required one to avoid Siegel zeroes, which restricted his results to a sequence of X going to infinity, rather than all sufficiently large X. However, it is possible to modify his argument to remove the effect of any exceptional zeroes, which allows us to extend the result to all sufficiently large X and also to make the implied constant effective. The intuitive reason for the $\frac{1}{k^2}$ factor is that our method produces, roughly speaking, k primes distributed "randomly" inside an interval of length about $\frac{\log X \log_2 X \log_4 X}{\log_3 X}$, and the narrowest gap between k independently chosen numbers in an interval of length L is typically of length about $\frac{1}{k^2} L$.

Our argument is based heavily on our previous paper [11], in particular using the hypergraph covering lemma from [11, Corollary 3] and the construction of sieve weights from [11, Theorem 5]. The main difference is in refining the probabilistic analysis in [11] to obtain good upper and lower bounds for certain sifted sets arising in the arguments in [11], whereas in the former paper only upper bounds were obtained.

We remark that in the recent paper [2], the methods from [11] were modified to obtain some information about the limit points of tuples of k consecutive prime gaps normalized by factors slightly slower than $\frac{\log X \log_2 X \log_4 X}{\log_3 X}$; see Theorem 6.4 of that paper for a precise statement.

1.1 Notational Conventions

In most of the paper, x will denote an asymptotic parameter going to infinity, with many quantities allowed to depend on x. The symbol $o(1)$ will stand for a quantity bounded in magnitude by $c(x)$, where $c(x)$ is a quantity that tends to zero as $x \to \infty$. The same convention applies to the asymptotic notation $X \sim Y$, which means $X = (1 + o(1))Y$, and $X \lesssim Y$, which means $X \leqslant (1 + o(1))Y$. We use $X = O(Y)$, $X \ll Y$, and $Y \gg X$ to denote the claim that there is a constant $C > 0$ such that $|X| \leqslant CY$ throughout the domain of the quantity X. We adopt the convention that C is independent of any parameter unless such dependence is indicated, e.g. by subscript such as $\ll_k$. In all of our estimates here, the constant C will be effective (we will not rely on ineffective results such as Siegel's theorem). If we can take the implied constant C to equal 1, we write $f = O_{\leqslant}(g)$ instead. Thus, for instance,

$$X = (1 + O_{\leqslant}(\varepsilon))Y$$

is synonymous with

$$(1 - \varepsilon)Y \leqslant X \leqslant (1 + \varepsilon)Y.$$

Finally, we use $X \asymp Y$ synonymously with $X \ll Y \ll X$.

When summing or taking products over the symbol p, it is understood that p is restricted to be prime.

Given a modulus q and an integer n, we use $n \bmod q$ to denote the congruence class of n in $\mathbb{Z}/q\mathbb{Z}$.

Given a set A, we use 1_A to denote its indicator function, thus $1_A(x)$ is equal to 1 when $x \in A$ and zero otherwise. Similarly, if E is an event or statement, we use 1_E to denote the indicator, equal to 1 when E is true and 0 otherwise. Thus, for instance, $1_A(x)$ is synonymous with $1_{x \in A}$.

We use $\#A$ to denote the cardinality of A, and for any positive real z, we let $[z] := \{n \in \mathbf{N} : 1 \leqslant n \leqslant z\}$ denote the set of natural numbers up to z.

Our arguments will rely heavily on the probabilistic method. Our random variables will mostly be discrete (in the sense that they take at most countably many values), although we will occasionally use some continuous random variables (e.g., independent real numbers sampled uniformly from the unit interval $[0, 1]$). As such, the usual measure-theoretic caveats such as "absolutely integrable," "measurable," or "almost surely" can be largely ignored by the reader in the discussion below. We will use boldface symbols such as $\mathbf{X}$ or $\mathbf{a}$ to denote random variables (and non-boldface symbols such as X or a to denote deterministic counterparts of these variables). Vector-valued random variables will be denoted in arrowed boldface, e.g. $\mathbf{a} = (\mathbf{a}_p)_{p \in \mathscr{P}}$ might denote a random tuple of random variables $\mathbf{a}_p$ indexed by some index set $\mathscr{P}$.

We write $\mathbb{P}$ for probability, and $\mathbb{E}$ for expectation. If $\mathbf{X}$ takes at most countably many values, we define the *essential range* of $\mathbf{X}$ to be the set of all X such that $\mathbb{P}(\mathbf{X} = X)$ is non-zero, thus $\mathbf{X}$ almost surely takes values in its essential range. We also employ the following conditional expectation notation. If E is an event of non-zero probability, we write

$$\mathbb{P}(F|E) := \frac{\mathbb{P}(F \wedge E)}{\mathbb{P}(E)}$$

for any event F, and

$$\mathbb{E}(\mathbf{X}|E) := \frac{\mathbb{E}(\mathbf{X}1_E)}{\mathbb{P}(E)}$$

for any (absolutely integrable) real-valued random variable $\mathbf{X}$. If $\mathbf{Y}$ is another random variable taking at most countably many values, we define the conditional probability $\mathbb{P}(F|\mathbf{Y})$ to be the random variable that equals $\mathbb{P}(F|\mathbf{Y} = Y)$ on the event $\mathbf{Y} = Y$ for each Y in the essential range of $\mathbf{Y}$, and similarly define the conditional expectation $\mathbb{E}(\mathbf{X}|\mathbf{Y})$ to be the random variable that equals $\mathbb{E}(\mathbf{X}|\mathbf{Y} = Y)$ on the event

$\mathbf{Y} = Y$. We observe the idempotency property

$$\mathbb{E}(\mathbb{E}(\mathbf{X}|\mathbf{Y})) = \mathbb{E}\mathbf{X} \tag{1}$$

whenever $\mathbf{X}$ is absolutely integrable and $\mathbf{Y}$ takes at most countably many values.

We will rely frequently on the following simple concentration of measure result.

Lemma 1 (Chebyshev Inequality) *Let X be a random variable with mean μ and let $\lambda > 0$. Then*

$$\mathbb{E}(|X - \mu| \geq \lambda) \leq \frac{\mathbb{E}X^2 - \mu^2}{\lambda^2}.$$

2   Siegel Zeroes

As is common in analytic number theory, we will have to address the possibility of an exceptional Siegel zero. As we want to keep all our estimates effective, we will not rely on Siegel's theorem or its consequences (such as the Bombieri-Vinogradov theorem). Instead, we will rely on the Landau-Page theorem, which we now recall. Throughout, χ denotes a Dirichlet character.

Lemma 2 (Landau-Page Theorem) *Let $Q \geqslant 100$. Suppose that $L(s, \chi) = 0$ for some primitive character χ of modulus at most Q, and some $s = \sigma + it$. Then either*

$$1 - \sigma \gg \frac{1}{\log(Q(1 + |t|))},$$

or else $t = 0$ and χ is a quadratic character χ_Q, which is unique for any given Q. Furthermore, if χ_Q exists, then its conductor q_Q is square-free apart from a factor of at most 4, and obeys the lower bound

$$q_Q \gg \frac{\log^2 Q}{\log_2^2 Q}.$$

Proof See, e.g., [7, Chapter 14]. The final estimate follows from the classical bound $1 - \beta \gg q^{-1/2} \log^{-2} q$ for a real zero β of $L(s, \chi)$ with χ of modulus q.

We can then eliminate the exceptional character by deleting at most one prime factor of Q.

Corollary 1 *Let $Q \geqslant 100$. Then there exists a quantity B_Q which is either equal to 1 or is a prime of size*

$$B_Q \gg \log_2 Q$$

with the property that

$$1 - \sigma \gg \frac{1}{\log(Q(1 + |t|))}$$

whenever $L(\sigma + it, \chi) = 0$ and χ is a character of modulus at most Q and coprime to B_Q.

Proof If the exceptional character χ_Q from Lemma 2 does not exist, then take $B_Q := 1$; otherwise, we take B_Q to be the largest prime factor of q_Q. As q_Q is square-free apart from a factor of at most 4, we have $\log q_Q \ll B_Q$ by the prime number theorem, and the claim follows.

Next, we recall Gallagher's prime number theorem:

Lemma 3 (Gallagher's Prime Number Theorem) *Let q be a natural number, and suppose that $L(s, \chi) \neq 0$ for all characters χ of modulus q and s with $1 - \sigma \leqslant \frac{\delta}{\log(Q(1+it))}$, and some constant $\delta > 0$. Then there is a constant $D \geqslant 1$ depending only on δ such that*

$$\#\{p \text{ prime} : p \leqslant x;\ p \equiv a \ (mod\ q)\} \gg \frac{x}{\phi(q) \log x}$$

for all $(a, q) = 1$ and $x \geqslant q^D$.

Proof See [14, Lemma 2] and [12].

This will combine well with Corollary 1 once we remove the moduli divisible by the (possible) exceptional prime B_Q.

3   Sieving an Interval

We now give the key sieving result that will be used to prove Theorem 1.

Theorem 2 (Sieving an Interval) *There is an absolute constant $c > 0$ such that the following holds. Fix $A \geqslant 1$ and $\varepsilon > 0$, and let x be sufficiently large depending on A and ε. Suppose y satisfies*

$$y = c \frac{x \log x \log_3 x}{\log_2 x}, \tag{2}$$

and suppose that $B_0 = 1$ or that B_0 is a prime satisfying

$$\log x \ll B_0 \leqslant x.$$

Then one can find a congruence class a_p mod p for each prime $p \leqslant x$, $p \neq B_0$ such that the sieved set

$$\mathscr{T} := \{n \in [y] \backslash [x] : n \not\equiv a_p \ (mod \ p) \ for \ all \ p \leqslant x, p \neq B_0\}$$

obeys the following size estimates:

- *(Upper Bound) One has*

$$\#\mathscr{T} \ll A \frac{x}{\log x}. \tag{3}$$

- *(Lower Bound) One has*

$$\#\mathscr{T} \gg A \frac{x}{\log x}. \tag{4}$$

- *(Upper bound in short intervals) For any $0 \leqslant \alpha \leqslant \beta \leqslant 1$, one has*

$$\#(\mathscr{T} \cap [\alpha y, \beta y]) \ll A(|\beta - \alpha| + \varepsilon) \frac{x}{\log x}. \tag{5}$$

We remark that if one lowers y to be of order $\frac{x \log x \log_3 x}{(\log_2 x)^2}$ rather than $\frac{x \log x \log_3 x}{\log_2 x}$, then this theorem is essentially [14, Lemma 6]. It is convenient to sieve $[y]\backslash[x]$ instead of $[y]$ for minor technical reasons (we will use the fact that the residue class 0 mod p avoids all the primes in $[y]\backslash[x]$ whenever $p \leqslant x$). The arguments in [11] already can give much of this theorem, with the exception of the lower bound (4), which is the main additional technical result of this paper that is needed to extend the results of that paper to longer chains.

We will prove Theorem 2 in later sections. In this section, we show how this theorem implies Theorem 1. Here we shall use the Maier matrix method, following the arguments in [14] closely (although we will use probabilistic notation rather than matrix notation). Let $k \geqslant 1$ be a fixed integer, let $c_0 > 0$ be a small constant, and let $A \geqslant 1$ and $0 < \varepsilon < 1/2$ be large and small quantities depending on k to be chosen later.

We now recall (a slight variant of) some lemmas from [14].

Lemma 4 *There exists an absolute constant $D \geqslant 1$ such that, for all sufficiently large x, there exists a natural number B_0 which is either equal to 1 or a prime, with*

$$\log x \ll B_0 \leqslant x, \tag{6}$$

and is such that the following holds. If one sets $P := P(x)/B_0$ (where we recall that $P(x)$ is the product of the primes up to x), then one has

$$\#\{z \in [Z] : Pz + a \text{ prime}\} \gg \frac{\log x}{\log Z} Z \tag{7}$$

for all $Z \geqslant P^D$ *and* $a \in P$ *coprime to* P, *and*

$$\#\{z \in [Z] : Pz + a, Pz + b \text{ both prime}\} \ll \left(\frac{\log x}{\log Z}\right)^2 Z \tag{8}$$

for all $Z \geqslant P^D$ *and all distinct* $a, b \in [P]$ *coprime to* P.

Proof We first prove (7). We apply Corollary 1 with $Q := P(x)$ to obtain a quantity $B_{P(x)}$ with the stated properties. We set $B_0 = 1$ if $B_{P(x)} > x$, and $B_0 := B_{P(x)}$ otherwise. Then from Mertens' theorem we have (6) if $B_0 \neq 1$. From Corollary 1 and Lemma 3, we then have

$$\#\{z \in [Z] : Pz + a \text{ prime}\} \gg \frac{PZ}{\phi(P)\log(PZ)}$$

for any $Z \geqslant P^D$ and a suitable absolute constant $D \geqslant 1$. Note that $\log(PZ) \ll \log Z$. From Mertens' theorem (and (6)) we also have

$$\frac{P}{\phi(P)} \asymp \log x, \tag{9}$$

and (7) follows.

Finally, the estimate (8) follows from standard upper bound sieves (cf. [14, Lemma 3]).

Now set $Z := P^D$ with x and D as in Lemma 4, and let $\mathbf{z}$ be chosen uniformly at random from $[Z]$. Let y, $\mathscr{T}$ and a_p mod p be as in Theorem 2. By the Chinese remainder theorem, we may find $m \in [P]$ such that $m \equiv -a_p \pmod{p}$ for all $p \leqslant x$ with $p \neq B_0$. Thus, $\mathbf{z}P + m + \mathscr{T}$ consists precisely of those elements of $\mathbf{z}P + m + [y]\backslash[x]$ that are coprime to P. In particular, any primes that lie in the interval $\mathbf{z}P + m + [y]\backslash[x]$ lie in $\mathbf{z}P + m + \mathscr{T}$.

From (7) and Mertens' theorem we have

$$\mathbb{P}(\mathbf{z}P + m + a \text{ prime}) \gg \frac{\log x}{x}$$

for all $a \in \mathscr{T}$ (we allow implied constants to depend on D). Similarly, from (8) and Mertens' theorem we have

$$\mathbb{P}(\mathbf{z}P + m + a, \mathbf{z}P(x) + m + b \text{ both prime}) \ll \left(\frac{\log x}{x}\right)^2 \tag{10}$$

for any distinct $a, b \in \mathscr{T}$. If we let $\mathbf{N}$ denote the number of primes in $\mathbf{z}P + m + \mathscr{T}$ (or equivalently, in $\mathbf{z}P + m + [y]\backslash[x]$), we thus have from (3) and (4) that

$$\mathbb{E}\mathbf{N} \gg A$$

and

$$\mathbb{E}\mathbf{N}^2 \ll A^2.$$

From this we see that with probability $\gg 1$, we have

$$A \ll \mathbf{N} \ll A, \tag{11}$$

where all implied constants are independent of ε and A. (This is because the contribution to $\mathbb{E}\mathbf{N}$ when $\mathbf{N}$ is much larger than A is much smaller than A.)

Next, if $0 \leqslant \alpha \leqslant \beta \leqslant 1$ and $\beta - \alpha \leqslant 2\varepsilon$, then from (10), (5) and the union bound we see that the probability that there are at least two primes in $\mathbf{z}P + m + [\alpha y, \beta y]$ is at most

$$O\left(\left(A\varepsilon \frac{x}{\log x}\right)^2 \left(\frac{\log x}{x}\right)^2\right) = O(A^2 \varepsilon^2).$$

Note that one can cover $[0, 1]$ with $O(1/\varepsilon)$ intervals of length at most 2ε, with the property that any two elements a, b of $[0, 1]$ with $|a - b| \leqslant \varepsilon$ may be covered by at least one of these intervals. From this and the union bound, we see that the probability that $\mathbf{z}P + m + [y]\backslash[x]$ contains two primes separated by at most εy is bounded by $O(\frac{1}{\varepsilon}A^2\varepsilon^2) = O(A^2\varepsilon)$. In particular, if we choose ε to be a sufficiently small multiple of $\frac{1}{A^2}$, we may find $z \in [Z]$ such that the interval $zP + m + [y]\backslash[x]$ contains $\gg A$ primes and has no prime gap less than εy. If we choose A to be a sufficiently large multiple of k, we conclude that

$$G_k(ZP + m + y) \geqslant \varepsilon y \gg \frac{1}{k^2} y.$$

By Mertens' theorem, we have $ZP + m + y \ll \exp(O(x))$, and Theorem 1 then follows from (2).

It remains to prove Theorem 2. This is the objective of the remaining sections of the paper.

4 Sieving a Set of Primes

Theorem 2 concerns the problem of deterministically sieving an interval $[y]\backslash[x]$ of size (2) so that the sifted set $\mathscr{T}$ has certain size properties. We use a variant of the Erdős-Rankin method to reduce this problem to a problem of *probabilistically sieving* a set $\mathscr{Q}$ of *primes* in $[y]\backslash[x]$, rather than integers in $[y]\backslash[x]$.

Given a real number $x \geqslant 1$, and a natural number B_0, define

$$z := x^{\log_3 x/(4\log_2 x)}, \tag{12}$$

and introduce the three disjoint sets of primes

$$\mathscr{S} := \{s \text{ prime} : \log^{20} x < s \leqslant z; s \neq B_0\}, \tag{13}$$

$$\mathscr{P} := \{p \text{ prime} : x/2 < p \leqslant x; p \neq B_0\}, \tag{14}$$

$$\mathscr{Q} := \{q \text{ prime} : x < q \leqslant y; q \neq B_0\}. \tag{15}$$

For residue classes $\mathbf{a} = (a_s \bmod s)_{s \in S}$ and $\mathbf{n} = (n_p \bmod p)_{p \in \mathscr{P}}$, define the sifted sets

$$S(\mathbf{a}) := \{n \in \mathbb{Z} : n \not\equiv a_s \ (\text{m}od\ s) \text{ for all } s \in \mathscr{S}\}$$

and likewise

$$S(\mathbf{n}) := \{n \in \mathbb{Z} : n \not\equiv n_p \ (\text{m}od\ p) \text{ for all } p \in \mathscr{P}\}.$$

We reduce Theorem 2 to

Theorem 3 (Sieving Primes) *Let $A \geqslant 1$ be a real number, let x be sufficiently large depending on A, and suppose that y obeys (2). Let B_0 be a natural number. Then there is a quantity*

$$A' \asymp A, \tag{16}$$

and some way to choose the vectors $\mathbf{a} = (\mathbf{a}_s \bmod s)_{s \in \mathscr{S}}$ and $\mathbf{n} = (\mathbf{n}_p \bmod p)_{p \in \mathscr{P}}$ at random (not necessarily independent of each other), such that for any fixed $0 \leqslant \alpha < \beta \leqslant 1$ (independent of x), one has with probability $1 - o(1)$ that

$$\#(\mathscr{Q} \cap S(\mathbf{a}) \cap S(\mathbf{n}) \cap (\alpha y, \beta y]) \sim A'|\beta - \alpha|\frac{x}{\log x}. \tag{17}$$

The $o(1)$ decay rates in the probability error and implied in the $\sim$ notation are allowed to depend on A, α, β.

In [11, Theorem 2], a weaker version of this theorem was established in which B_0 was not present, and only the upper bound in (17) was proven. Thus, the main new contribution of this paper is the lower bound in (17).

We prove Theorem 3 in subsequent sections. In this section, we show how this theorem implies Theorem 2 (and hence Theorem 1). The arguments here are almost identical to those in [11, §2].

Fix $A \geqslant 1, 0 < \varepsilon \leqslant 1$. We partition $(0, 1]$ into $O(1/\varepsilon)$ intervals $[\alpha_i, \beta_i]$ of length between $\varepsilon/2$ and ε. Applying Theorem 3 with the pairs $(\alpha, \beta) = (\alpha_i, \beta_i)$ and the pair $(\alpha, \beta) = (0, 1)$, and invoking a union bound (and the fact that ε is independent of x), we see that if x is sufficiently large (depending on A, ε), there are A', y obeying (16), (2) and tuples of residue classes $\mathbf{a} = (a_s \bmod s)_{s \in \mathscr{S}}$ and $\mathbf{n} = (n_p \bmod p)_{p \in \mathscr{P}}$ such that

$$\#(\mathscr{Q} \cap S(\mathbf{a}) \cap S(\mathbf{n})) \sim A' \frac{x}{\log x}$$

and

$$\#(\mathscr{Q} \cap S(\mathbf{a}) \cap S(\mathbf{n})) \cap (\alpha_i y, \beta_i y]) \ll A\varepsilon \frac{x}{\log x}$$

for all i. A covering argument then gives

$$\#(\mathscr{Q} \cap S(\mathbf{a}) \cap S(\mathbf{n}) \cap [\alpha y, \beta y]) \ll A(|\beta - \alpha| + \varepsilon) \frac{x}{\log x}$$

for any $0 \leqslant \alpha < \beta \leqslant 1$. Now we extend the tuple $\mathbf{a}$ to a tuple $(a_p)_{p \leqslant x}$ of congruence classes $a_p \bmod p$ for all primes $p \leqslant x$ by setting $a_p := n_p$ for $p \in \mathscr{P}$ and $a_p := 0$ for $p \notin \mathscr{S} \cup \mathscr{P}$, and consider the sifted set

$$\mathscr{T} := \{n \in [y] \backslash [x] : n \not\equiv a_p \ (\mathrm{mod} \ p) \text{ for all } p \leqslant x\}.$$

The elements of $\mathscr{T}$, by construction, are not divisible by any prime in $(0, \log^{20} x]$ or in $(z, x/2]$, except possibly for B_0. Thus, each element must either be a z-smooth number (i.e., a number with all prime factors at most z) times a power of B_0, or must consist of a prime greater than $x/2$, possibly multiplied by some additional primes that are all either at least $\log^{20} x$ or equal to B_0. However, from (2) we know that $y = o(x \log x)$, and by hypothesis we know that $B_0 \gg \log x$. Thus, we see that an element of $\mathscr{T}$ is either a z-smooth number times a power of B_0 or a prime in $\mathscr{Q}$. In the second case, the element lies in $\mathscr{Q} \cap S(\mathbf{a}) \cap S(\mathbf{n})$. Conversely, every element of $\mathscr{Q} \cap S(\mathbf{a}) \cap S(\mathbf{n})$ lies in $\mathscr{T}$. Thus, $\mathscr{T}$ only differs from $\mathscr{Q} \cap S(\mathbf{a}) \cap S(\mathbf{n})$ by a set $\mathscr{R}$ consisting of z-smooth numbers in $[y]$ multiplied by powers of B_0.

To estimate $\#\mathscr{R}$, let

$$u := \frac{\log y}{\log z},$$

so from (2), (12) one has $u \sim 4 \frac{\log_2 x}{\log_3 x}$. The number of powers of B_0 in $[y]$ is $O(\log x)$. By standard counts for smooth numbers (e.g., de Bruijn's theorem [8]) and (2), we thus have

$$\#\mathscr{R} \ll \log x \times y e^{-u \log u + O(u \log \log (u+2))}$$

$$= \log x \times \frac{y}{\log^{4+o(1)} x} = o\left(\frac{x}{\log x}\right).$$

Thus the contribution of $\mathscr{R}$ to $\mathscr{T}$ is negligible for the purposes of establishing the bounds (3)–(5), and Theorem 2 follows from (17).

It remains to establish Theorem 3. This is the objective of the remaining sections of the paper.

5   Using a Hypergraph Covering Theorem

In the previous section we reduced matters to obtaining random residue classes $\mathbf{a}, \mathbf{n}$ such that the sifted set $\mathscr{Q} \cap S(\mathbf{a}) \cap S(\mathbf{n})$ is small. In this section we use a hypergraph covering theorem from [11] to reduce the task to that of finding random residue classes $\mathbf{n}$ that have large intersection with $\mathscr{Q} \cap S(\mathbf{a})$. More precisely, we will use the following result:

Theorem 4   *Let $x \to \infty$. Let $\mathscr{P}'$, $\mathscr{Q}'$ be sets of primes in $(x/2, x]$ and $(x, x \log x]$, respectively, with $\#\mathscr{Q}' > (\log_2 x)^3$. For each $p \in \mathscr{P}'$, let $\mathbf{e}_p$ be a random subset of $\mathscr{Q}'$ satisfying the size bound*

$$\#\mathbf{e}_p \leqslant r = O\left(\frac{\log x \log_3 x}{\log_2^2 x}\right) \qquad (p \in \mathscr{P}'). \tag{18}$$

Assume the following:

- *(Sparsity) For all $p \in \mathscr{P}'$ and $q \in \mathscr{Q}'$,*

$$\mathbb{P}(q \in \mathbf{e}_p) \leqslant x^{-1/2-1/10}. \tag{19}$$

- *(Uniform covering) For all but at most $\frac{1}{(\log_2 x)^2}\#\mathscr{Q}'$ elements $q \in \mathscr{Q}'$, we have*

$$\sum_{p \in \mathscr{P}'} \mathbb{P}(q \in \mathbf{e}_p) = C + O_{\leqslant}\left(\frac{1}{(\log_2 x)^2}\right) \tag{20}$$

  for some quantity C, independent of q, satisfying

$$\frac{5}{4} \log 5 \leqslant C \ll 1. \tag{21}$$

- *(Small codegrees) For any distinct $q_1, q_2 \in \mathscr{Q}'$,*

$$\sum_{p \in \mathscr{P}'} \mathbb{P}(q_1, q_2 \in \mathbf{e}_p) \leqslant x^{-1/20}. \tag{22}$$

Then for any positive integer m with

$$m \leqslant \frac{\log_3 x}{\log 5}, \tag{23}$$

we can find random sets $\mathbf{e}'_p \subseteq \mathcal{Q}'$ *for each* $p \in \mathcal{P}'$ *such that*

$$\#\{q \in \mathcal{Q}' : q \notin \mathbf{e}'_p \text{ for all } p \in \mathcal{P}'\} \sim 5^{-m} \# \mathcal{Q}'$$

with probability $1 - o(1)$. *More generally, for any* $\mathcal{Q}'' \subset \mathcal{Q}'$ *with cardinality at least* $(\#\mathcal{Q}')/\sqrt{\log_2 x}$, *one has*

$$\#\{q \in \mathcal{Q}'' : q \notin \mathbf{e}'_p \text{ for all } p \in \mathcal{P}'\} \sim 5^{-m} \# \mathcal{Q}''$$

with probability $1 - o(1)$. *The decay rates in the* $o(1)$ *and* $\sim$ *notation are uniform in* $\mathcal{P}'$, $\mathcal{Q}'$, $\mathcal{Q}''$.

Proof See [11, Corollary 3]. $\qquad \blacksquare$

In view of the above result, we may now reduce Theorem 3 to the following claim.

Theorem 5 (Random Construction) *Let x be a sufficiently large real number, let* B_0 *be a natural number and suppose y satisfies* (2). *Then there is a quantity C with*

$$C \asymp \frac{1}{c} \tag{24}$$

with the implied constants independent of c, and some way to choose random vectors $\mathbf{a} = (\mathbf{a}_s \bmod s)_{s \in \mathcal{S}}$ *and* $\mathbf{n} = (\mathbf{n}_p)_{p \in \mathcal{P}}$ *of congruence classes* $\mathbf{a}_s$ mod s *and integers* $\mathbf{n}_p$, *obeying the following axioms:*

- *For every* a *in the essential range of* $\mathbf{a}$, *one has*

$$\mathbb{P}(q \equiv \mathbf{n}_p \ (mod \ p)|\mathbf{a} = a) \leqslant x^{-1/2 - 1/10}$$

  uniformly for all $p \in \mathcal{P}$.
- *For fixed* $0 \leqslant \alpha < \beta \leqslant 1$, *we have with probability* $1 - o(1)$ *that*

$$\#(\mathcal{Q} \cap S(a) \cap [\alpha y, \beta y]) \sim 80c|\beta - \alpha| \frac{x}{\log x} \log_2 x. \tag{25}$$

- *Call an element* a *in the essential range of* $\mathbf{a}$ *good if, for all but at most* $\frac{x}{\log x \log_2 x}$ *elements* $q \in \mathcal{Q} \cap S(a)$, *one has*

$$\sum_{p \in \mathcal{P}} \mathbb{P}(q \equiv \mathbf{n}_p \ (mod \ p)|\mathbf{a} = a) = C + O_{\leqslant}\left(\frac{1}{(\log_2 x)^2}\right). \tag{26}$$

Then **a** *is good with probability* $1 - o(1)$.

We now show why Theorem 5 implies Theorem 3. By (24), we may choose $0 < c < 1/2$ small enough so that (21) holds. Let $A \geqslant 1$ be a fixed quantity. Then we can find an integer m obeying (23) such that the quantity

$$A' := 5^{-m} \times 80c \log_2 x$$

is such that $A' \asymp A$ with implied constants independent of A.

Suppose that we are in the probability $1 - o(1)$ event that **a** takes a value **a** which is good and such that (25) holds. On each sub-event $\mathbf{a} = \mathbf{a}$ of this probability $1 - o(1)$ event, we may apply Theorem 4 (for the random variables $\mathbf{n}_p$ conditioned to this event) define the random variables $\mathbf{n}'_p$ on this event with the stated properties. For the remaining events $\mathbf{a} = \mathbf{a}$, we set $\mathbf{n}'_p$ arbitrarily (e.g., we could set $\mathbf{n}'_p = 0$). The claim (17) then follows from Theorem 4 and (25), thus establishing Theorem 3.

It remains to establish Theorem 5. This will be achieved in the next section.

6   Using a Sieve Weight

If r is a natural number, an *admissible r-tuple* is a tuple $(h_1, \ldots, h_r)$ of distinct integers $h_1, \ldots, h_r$ that do not cover all residue classes modulo p, for any prime p. For instance, the tuple $(p_{\pi(r)+1}, \ldots, p_{\pi(r)+r})$ consisting of the first r primes larger than r is an admissible r-tuple.

We will establish Theorem 5 by a probabilistic argument involving a certain weight function. More precisely, we will deduce this result from the following construction from [11].

Theorem 6 (Existence of Good Sieve Weight) *Let x be a sufficiently large real number, let B_0 be an integer, and let y be any quantity obeying (2). Let $\mathscr{P}, \mathscr{Q}$ be defined by (14), (15). Let r be a positive integer with*

$$r_0 \leqslant r \leqslant \log^{c_0} x \tag{27}$$

for some sufficiently small absolute constant c_0 and sufficiently large absolute constant r_0, and let $(h_1, \ldots, h_r)$ be an admissible r-tuple contained in $[2r^2]$. Then one can find a positive quantity

$$\tau \geqslant x^{-o(1)} \tag{28}$$

and a positive quantity $u = u(r)$ depending only on r with

$$u \asymp \log r \tag{29}$$

and a non-negative function $w : \mathscr{P} \times \mathbb{Z} \to \mathbb{R}^+$ supported on $\mathscr{P} \times (\mathbb{Z} \cap [-y, y])$ with the following properties:

- *Uniformly for every $p \in \mathscr{P}$, one has*

$$\sum_{n \in \mathbb{Z}} w(p, n) = \left(1 + O\left(\frac{1}{\log_2^{10} x}\right)\right) \tau \frac{y}{\log^r x}. \tag{30}$$

- *Uniformly for every $q \in \mathscr{Q}$ and $i = 1, \ldots, r$, one has*

$$\sum_{p \in \mathscr{P}} w(p, q - h_i p) = \left(1 + O\left(\frac{1}{\log_2^{10} x}\right)\right) \tau \frac{u}{r} \frac{x}{2 \log^r x}. \tag{31}$$

- *Uniformly for every $h = O(y/x)$ that is not equal to any of the h_i, one has*

$$\sum_{q \in \mathscr{Q}} \sum_{p \in \mathscr{P}} w(p, q - hp) = O\left(\frac{1}{\log_2^{10} x} \tau \frac{x}{\log^r x} \frac{y}{\log x}\right). \tag{32}$$

- *Uniformly for all $p \in \mathscr{P}$ and $n \in \mathbb{Z}$,*

$$w(p, n) = O(x^{1/3 + o(1)}). \tag{33}$$

Proof See[2] [11, Theorem 5]. We remark that the construction of the weights and the verification of the required estimates relies heavily on the previous work of the second author in [16]. $\square$

It remains to show how Theorem 6 implies Theorem 5. The analysis will be based on that in [11, §5], which used a weight with slightly weaker hypotheses than in Theorem 6 to obtain somewhat weaker conclusions than Theorem 5 (in which the condition $q \equiv \mathbf{n}_p \pmod{p}$ was replaced by the stronger condition that $q = \mathbf{n}_p + h_i p$ for some $i = 1, \ldots, r$).

Let $x, B_0, c, y, z, \mathscr{S}, \mathscr{P}, \mathscr{Q}$ be as in Theorem 5. Let c_0 be a sufficiently small absolute constant. We set r to be the maximum value permitted by Theorem 6, namely

$$r := \lfloor \log^{c_0} x \rfloor \tag{34}$$

and let $(h_1, \ldots, h_r)$ be the admissible r-tuple consisting of the first r primes larger than r, thus $h_i = p_{\pi(r)+i}$ for $i = 1, \ldots, r$. From the prime number theorem we have $h_i = O(r \log r)$ for $i = 1, \ldots, r$, and so we have $h_i \in [2r^2]$ for $i = 1, \ldots, r$ if x is large enough (there are many other choices possible, e.g. $(h_1, \ldots, h_r) = (1^2, 3^2, \ldots, (2r - 1)^2)$). We now invoke Theorem 6 to obtain quantities τ, u and a weight $w : \mathscr{P} \times \mathbb{Z} \to \mathbb{R}^+$ with the stated properties.

For each $p \in \mathscr{P}$, let $\tilde{\mathbf{n}}_p$ denote the random integer with probability density

---

[2]The integer B_0 was not deleted from the sets $\mathscr{P}$ or $\mathscr{Q}$ in that theorem, however it is easy to see (using (33)) that deleting at most one prime from either $\mathscr{P}$ or $\mathscr{Q}$ will not significantly worsen any of the estimates claimed by the theorem.

$$\mathbb{P}(\tilde{\mathbf{n}}_p = n) := \frac{w(p, n)}{\sum_{n' \in \mathbb{Z}} w(p, n')}$$

for all $n \in \mathbb{Z}$ (we will not need to impose any independence conditions on the $\tilde{\mathbf{n}}_p$). From (30), (31) we have

$$\sum_{p \in \mathscr{P}} \mathbb{P}(q = \tilde{\mathbf{n}}_p + h_i\, p) = \left(1 + O\left(\frac{1}{\log_2^{10} x}\right)\right) \frac{u}{r} \frac{x}{2y} \tag{35}$$

for every $q \in \mathscr{Q}$ and $i = 1, \dots, r$, and similarly from (30), (32) we have

$$\sum_{q \in \mathscr{Q}} \sum_{p \in \mathscr{P}} \mathbb{P}(q = \tilde{\mathbf{n}}_p + hp) \ll \frac{1}{\log_2^{10} x} \frac{x}{\log x} \tag{36}$$

for every $h = O(y/x)$ not equal to any of the h_i. Finally, from (30), (33), (28) one has

$$\mathbb{P}(\tilde{\mathbf{n}}_p = n) \ll x^{-1/2 - 1/6 + o(1)} \tag{37}$$

for all $p \in \mathscr{P}$ and $n \in \mathbb{Z}$.

We choose the random vector $\mathbf{a} := (\mathbf{a}_s \bmod s)_{s \in \mathscr{S}}$ by selecting each $\mathbf{a}_s \bmod s$ uniformly at random from $\mathbb{Z}/s\mathbb{Z}$, independently in s and independently of the $\tilde{\mathbf{n}}_p$. The resulting sifted set $S(\mathbf{a})$ is a random periodic subset of $\mathbb{Z}$ with density

$$\sigma := \prod_{s \in \mathscr{S}} \left(1 - \frac{1}{s}\right).$$

From the prime number theorem (with sufficiently strong error term), (12) and (13),

$$\sigma = \left(1 + O\left(\frac{1}{\log_2^{10} x}\right)\right) \frac{\log(\log^{20} x)}{\log z} = \left(1 + O\left(\frac{1}{\log_2^{10} x}\right)\right) \frac{80 \log_2 x}{\log x \, \log_3 x / \log_2 x},$$

so in particular we see from (2) that

$$\sigma y = \left(1 + O\left(\frac{1}{\log_2^{10} x}\right)\right) 80cx \log_2 x. \tag{38}$$

We also see from (34) that

$$\sigma^r = x^{o(1)}. \tag{39}$$

We have a useful correlation bound:

Lemma 5 *Let $t \leqslant \log x$ be a natural number, and let $n_1, \ldots, n_t$ be distinct integers of magnitude $O(x^{O(1)})$. Then one has*

$$\mathbb{P}(n_1, \ldots, n_t \in S(\mathbf{a})) = \left(1 + O\left(\frac{1}{\log^{16} x}\right)\right) \sigma^t.$$

Proof See [11, Lemma 5.1].

Among other things, this gives the claim (25):

Corollary 2 *For any fixed $0 \leqslant \alpha < \beta \leqslant 1$, we have with probability $1 - o(1)$ that*

$$\#(\mathscr{Q} \cap [\alpha y, \beta y] \cap S(\mathbf{a})) \sim \sigma|\beta - \alpha|\frac{y}{\log x} \sim 80c|\beta - \alpha|\frac{x}{\log x}\log_2 x. \qquad (40)$$

Proof See [11, Corollary 4], replacing $\mathscr{Q}$ with $\mathscr{Q} \cap [\alpha y, \beta y]$.

For each $p \in \mathscr{P}$, we consider the quantity

$$X_p(\mathbf{a}) := \mathbb{P}(\tilde{\mathbf{n}}_p + h_i p \in S(\mathbf{a}) \text{ for all } i = 1, \ldots, r), \qquad (41)$$

and let $\mathscr{P}(\mathbf{a})$ denote the set of all the primes $p \in \mathscr{P}$ such that

$$X_p(\mathbf{a}) = \left(1 + O_{\leqslant}\left(\frac{1}{\log^3 x}\right)\right) \sigma^r. \qquad (42)$$

In light of Lemma 5, we expect most primes in $\mathscr{P}$ to lie in $\mathscr{P}(\mathbf{a})$, and this will be confirmed below (Lemma 6). We now define the random variables $\mathbf{n}_p$ as follows. Suppose we are in the event $\mathbf{a} = \mathbf{a}$ for some $\mathbf{a}$ in the range of $\mathbf{a}$. If $p \in \mathscr{P} \backslash \mathscr{P}(\mathbf{a})$, we set $\mathbf{n}_p = 0$. Otherwise, if $p \in \mathscr{P}(\mathbf{a})$, we define $\mathbf{n}_p$ to be the random integer with conditional probability distribution

$$\mathbb{P}(\mathbf{n}_p = n | \mathbf{a} = \mathbf{a}) := \frac{Z_p(\mathbf{a}; n)}{X_p(\mathbf{a})}, \quad Z_p(\mathbf{a}; n) = 1_{n + h_j p \in S(\mathbf{a}) \text{ for } j = 1, \ldots, r} \mathbb{P}(\tilde{\mathbf{n}}_p = n). \qquad (43)$$

with the $\mathbf{n}_p$ jointly conditionally independent on the event $\mathbf{a} = \mathbf{a}$. From (42) we see that these random variables are well defined.

Substituting definition (43) into the left-hand side of (26), and observing that $\mathbf{n}_p \equiv q \pmod{p}$ is only possible if $p \in \mathscr{P}(\mathbf{a})$, we see that to prove (26), it suffices to show that with probability $1 - o(1)$ in $\mathbf{a}$, for all but at most $\frac{x}{\log x \log_2 x}$ primes in $\mathscr{Q} \cap S(\mathbf{a})$, we have

$$\sigma^{-r} \sum_{p \in \mathscr{P}(\mathbf{a})} \sum_h Z_p(\mathbf{a}; q - hp) = C + O\left(\frac{1}{\log_2^3 x}\right). \qquad (44)$$

We now confirm that $\mathscr{P} \backslash \mathscr{P}(\mathbf{a})$ is small with high probability.

Lemma 6 *With probability $1 - O(1/\log^3 x)$, $\mathscr{P}(\mathbf{a})$ contains all but $O(\frac{1}{\log^3 x} \frac{x}{\log x})$ of the primes $p \in \mathscr{P}$. In particular, $\mathbb{E}\#\mathscr{P}(\mathbf{a}) = \#\mathscr{P}(1 + O(1/\log^3 x))$.*

Proof See [11, Lemma 5.3].

The left side of relation (44) breaks naturally into two pieces, a "main term" consisting of summands where $h = h_i$ for some i, and an "error terms" consisting of the remaining summands. We first take care of the error terms.

Lemma 7 *With probability $1 - o(1)$ we have*

$$\sigma^{-r} \sum_{\substack{p \in \mathscr{P}(\mathbf{a})}} \sum_{\substack{h \ll y/x \\ h \notin \{h_1, \ldots, h_r\}}} Z_p(\mathbf{a}; q - hp) \ll \frac{1}{\log_2^3 x} \tag{45}$$

for all but at most $\frac{x}{2 \log x \log_2 x}$ primes $q \in \mathscr{Q} \cap S(\mathbf{a})$.

Proof We first extend the sum over all $p \in \mathscr{P}$. By Markov's inequality, it suffices to show that

$$\mathbb{E} \sum_{\substack{q \in \mathscr{Q} \cap S(\mathbf{a})}} \sigma^{-r} \sum_{\substack{p \in \mathscr{P}}} \sum_{\substack{h \ll y/x \\ h \notin \{h_1, \ldots, h_k\}}} Z_p(\mathbf{a}; q - hp) = o\left(\frac{x}{\log x \log_2^4 x}\right). \tag{46}$$

The left-hand side of (46) equals

$$\sigma^{-r} \sum_{\substack{q \in \mathscr{Q}}} \sum_{\substack{h \ll y/x \\ h \notin \{h_1, \ldots, h_k\}}} \sum_{\substack{p \in \mathscr{P}}} \mathbb{P}(q \in S(\mathbf{a}), q + h_j p - hp \in S(\mathbf{a}) \text{ for } j = 1, \ldots, r)$$

$$\times \mathbb{P}(q = \tilde{\mathbf{n}}_p + hp).$$

We note that for any h in the above sum, the $r + 1$ integers $q, q + h_1 p - hp, \ldots, q + h_r p - hp$ are distinct. Applying Lemma 5, followed by (36), we may thus bound this expression by

$$\ll \sum_{\substack{h \ll y/x \\ h \notin \{h_1, \ldots, h_k\}}} \sigma \frac{x/\log x}{\log_2^{10} x} \ll \sigma \frac{1}{\log_2^{10} x} \frac{y}{\log x}.$$

The claim now follows from (38).

Next, we deal with the main term of (44), by showing an analogue of (35).

Lemma 8 *With probability $1 - o(1)$, we have*

$$\sigma^{-r} \sum_{i=1}^{r} \sum_{p \in \mathscr{P}(\mathbf{a})} Z_p(\mathbf{a}; q - h_i p) = \left(1 + O\left(\frac{1}{\log_2^3 x}\right)\right) \frac{u}{\sigma} \frac{x}{2y}. \tag{47}$$

for all but at most $\frac{x}{2 \log x \log_2 x}$ of the primes $q \in \mathscr{Q} \cap S(\mathbf{a})$.

Proof We first show that replacing $\mathscr{P}(\mathbf{a})$ with $\mathscr{P}$ has negligible effect on the sum, with probability $1 - o(1)$. Fix i and substitute $n = q - h_i p$. By Markov's inequality, it suffices to show that

$$\mathbb{E} \sum_n \sigma^{-r} \sum_{p \in \mathscr{P} \setminus \mathscr{P}(\mathbf{a})} Z_p(\mathbf{a}; n) = o\left(\frac{u}{\sigma} \frac{x}{2y} \frac{1}{r} \frac{1}{\log_2^3 x} \frac{x}{\log x \log_2 x}\right). \tag{48}$$

By Lemma 5, we have

$$\mathbb{E} \sum_n \sigma^{-r} \sum_{p \in \mathscr{P}} Z_p(\mathbf{a}; n) = \sigma^{-r} \sum_{p \in \mathscr{P}} \sum_n \mathbb{P}(\tilde{\mathbf{n}}_p = n)$$

$$\times \mathbb{P}(n + h_j p \in S(\mathbf{a}) \text{ for } j = 1, \ldots, r)$$

$$= \left(1 + O\left(\frac{1}{\log^{16} x}\right)\right) \#\mathscr{P}.$$

Next, by (42) and Lemma 6 we have

$$\mathbb{E} \sum_n \sigma^{-r} \sum_{p \in \mathscr{P}(\mathbf{a})} Z_p(\mathbf{a}; n) = \sigma^{-r} \sum_{\mathbf{a}} \mathbb{P}(\mathbf{a} = \mathbf{a}) \sum_{p \in \mathscr{P}(\mathbf{a})} X_p(\mathbf{a})$$

$$= \left(1 + O\left(\frac{1}{\log^3 x}\right)\right) \mathbb{E} \#\mathscr{P}(\mathbf{a}) = \left(1 + O\left(\frac{1}{\log^3 x}\right)\right) \#\mathscr{P};$$

subtracting, we conclude that the left-hand side of (48) is $O(\#\mathscr{P} / \log^3 x) = O(x / \log^4 x)$. The claim then follows from (2) and (27).

By (48), it suffices to show that with probability $1 - o(1)$, for all but at most $\frac{x}{2 \log x \log_2 x}$ primes $q \in \mathscr{Q} \cap S(\mathbf{a})$, one has

$$\sum_{i=1}^{r} \sum_{p \in \mathscr{P}} Z_p(\mathbf{a}; q - h_i p) = \left(1 + O_{\leqslant}\left(\frac{1}{\log_2^3 x}\right)\right) \sigma^{r-1} u \frac{x}{2y}. \tag{49}$$

Call a prime $q \in \mathscr{Q}$ *bad* if $q \in \mathscr{Q} \cap S(\mathbf{a})$ but (49) fails. Using Lemma 5 and (35), we have

$$\mathbb{E}\left[\sum_{q \in \mathscr{Q} \cap S(\mathbf{a})} \sum_{i=1}^{r} \sum_{p \in \mathscr{P}} Z_p(\mathbf{a}; q - h_i p)\right]$$

$$= \sum_{q,i,p} \mathbb{P}(q + (h_j - h_i)p \in S(\mathbf{a}) \text{ for all } j = 1, \ldots, r) \mathbb{P}(\tilde{\mathbf{n}}_p = q - h_i p)$$

$$= \left(1 + O\left(\frac{1}{\log_2^{10} x}\right)\right) \frac{\sigma y}{\log x} \sigma^{r-1} u \frac{x}{2y}$$

and

$$\mathbb{E}\left[\sum_{q \in \mathcal{Q} \cap S(\mathbf{a})} \left(\sum_{i=1}^{r} \sum_{p \in \mathscr{P}} Z_p(\mathbf{a}; q - h_i p)\right)^2\right]$$

$$= \sum_{\substack{p_1, p_2, q \\ i_1, i_2}} \mathbb{P}(q + (h_j - h_{i_\ell})p_\ell \in S(\mathbf{a}) \text{ for } j = 1, \ldots, r; \ell = 1, 2)$$

$$\times \mathbb{P}(\tilde{\mathbf{n}}_{p_1}^{(1)} = q - h_{i_1} p_1)\mathbb{P}(\tilde{\mathbf{n}}_{p_2}^{(2)} = q - h_{i_2} p_2)$$

$$= \left(1 + O\left(\frac{1}{\log_2^{10} x}\right)\right) \frac{\sigma y}{\log x} \left(\sigma^{r-1} u \frac{x}{2y}\right)^2,$$

where $(\tilde{\mathbf{n}}_{p_1}^{(1)})_{p_1 \in \mathscr{P}}$ and $(\tilde{\mathbf{n}}_{p_2}^{(2)})_{p_2 \in \mathscr{P}}$ are independent copies of $(\tilde{\mathbf{n}}_p)_{p \in \mathscr{P}}$ over $\mathbf{a}$. In the last step we used the fact that the terms with $p_1 = p_2$ contribute negligibly.

By Chebyshev's inequality (Lemma 1) it follows that the number of bad q is $\ll \frac{\sigma y}{\log x} \frac{1}{\log_2^3 x} \ll \frac{x}{\log x \log_2^2 x}$ with probability $1 - O(1/\log_2 x)$. This concludes the proof.

We now conclude the proof of Theorem 5. We need to prove (44); this follows immediately from Lemmas 7 and 8 upon noting that by (34), (29) and (38),

$$C := \frac{u}{\sigma} \frac{x}{2y} \sim \frac{1}{c}.$$

Acknowledgements Kevin Ford thanks the hospitality of the Institute of Mathematics and Informatics of the Bulgarian Academy of Sciences. The research of James Maynard was conducted partly while he was a CRM-ISM postdoctoral fellow at the Université de Montréal, and partly while he was a Fellow by Examination at Magdalen College, Oxford.

Kevin Ford was supported by NSF grants DMS-1201442 and DMS-1501982. Terence Tao was supported by a Simons Investigator grant, the James and Carol Collins Chair, the Mathematical Analysis & Application Research Fund Endowment, and by NSF grant DMS-1266164.

The authors thank Tristan Freiberg for some corrections.

References

1. R.J. Backlund, Über die Differenzen zwischen den Zahlen, die zu den ersten n Primzahlen teilerfremd sind. Ann. Acad. Sci. Fenn. **32**(2), 1–9 (1929). Commentationes in honorem E. L. Lindelöf

2. R.C. Baker, T. Freiberg, Limit points and long gaps between primes. Q. J. Math. Oxford **67**(2), 233–260 (2016)
3. R.C. Baker, G. Harman, J. Pintz, The difference between consecutive primes. II. Proc. Lond. Math. Soc. (3) **83**(3), 532–562 (2001)
4. A. Brauer, H. Zeitz, Über eine zahlentheoretische Behauptung von Legendre. Sber. Berliner Math. Ges. **29**, 116–125 (1930)
5. H. Cramér, Some theorems concerning prime numbers. Ark. Mat. Astr. Fys. **15**, 1–33 (1920)
6. H. Cramér, On the order of magnitude of the difference between consecutive prime numbers. Acta Arith. **2**, 396–403 (1936)
7. H. Davenport, *Multiplicative Number Theory*. Graduate Texts in Mathematics, 3rd edn., vol. 74 (Springer, New York, 2000)
8. N.G. de Bruijn, On the number of positive integers $\leqslant x$ and free of prime factors $> y$. Nederl. Acad. Wetensch. Proc. Ser. A. **54**, 50–60 (1951)
9. P. Erdős, On the difference of consecutive primes. Q. J. Math. Oxford Ser. **6**, 124–128 (1935)
10. K. Ford, B. Green, S. Konyagin, T. Tao, Large gaps between consecutive prime numbers. Ann. Math. **183**, 935–974 (2016)
11. K. Ford, B. Green, S. Konyagin, J. Maynard, T. Tao, Long gaps between primes. J. Am. Math. Soc. **31**(1), 65–105 (2018)
12. P.X. Gallagher, A large sieve density estimate near $\sigma = 1$. Invent. Math. **11**, 329–339 (1970)
13. A. Granville, Harald Cramér and the distribution of prime numbers. Scand. Actuar. J. **1**, 12–28 (1995)
14. H. Maier, Chains of large gaps between consecutive primes. Adv. Math. **39**, 257–269 (1981)
15. H. Maier, C. Pomerance, Unusually large gaps between consecutive primes. Trans. Am. Math. Soc. **322**(1), 201–237 (1990)
16. J. Maynard, Dense clusters of primes in subsets. Compos. Math. **152**(7), 1517–1554 (2016)
17. J. Maynard, Dense clusters of primes in subsets. Compos. Math. **152**(7), 1517–1554 (2016)
18. J. Pintz, On the distribution of gaps between consecutive primes. Preprint
19. J. Pintz, Very large gaps between consecutive primes. J. Number Theory **63**(2), 286–301 (1997)
20. R.A. Rankin, The difference between consecutive prime numbers. J. Lond. Math. Soc. **13**, 242–247 (1938)
21. R.A. Rankin, The difference between consecutive prime numbers. V. Proc. Edinb. Math. Soc. (2) **13**, 331–332 (1962/1963)
22. A. Schönhage, Eine Bemerkung zur Konstruktion grosser Primzahllücken. Arch. Math. **14**, 29–30 (1963)
23. E. Westzynthius, Über die Verteilung der Zahlen, die zu den n ersten Primzahlen teilerfremd sind. Societas Scientarium Fennica, Helsingfors **5**(25), 1–37 (1931). Commentationes Physico–Mathematicae

A Note on the Distribution of Primes in Intervals

Tristan Freiberg

Abstract Assuming a certain form of the Hardy–Littlewood prime tuples conjecture, we show that, given any positive numbers $\lambda_1, \ldots, \lambda_r$ and nonnegative integers $m_1, \ldots, m_r$, the proportion of positive integers $n \leqslant x$ for which, for each $j \leqslant r$, the interval $(n, n + (\lambda_1 + \cdots + \lambda_j) \log x]$ contains exactly $m_1 + \cdots + m_j$ primes, is asymptotically equal to $\prod_{j=1}^{r}(e^{-\lambda_j}\lambda^{m_j}/m_j!)$ as $x \to \infty$. This extends a result of Gallagher, who considered the case $r = 1$. We use a direct inclusion–exclusion argument in place of Gallagher's moment calculation, thereby avoiding recourse to the moment determinacy property of Poisson distributions.

1 Introduction

Irregularities in the distribution of primes are manifest in results that appear at odds with Cramér's model. Such results are exemplified in Maier's sensational paper [18], which offers a newfound understanding of the distribution of primes. The Hardy–Littlewood prime tuples conjecture is also apparently at odds with Cramér's model: it asserts (quantitatively) that, given a finite set of integers $\boldsymbol{h}$, the integers $n + h$, $h \in \boldsymbol{h}$, are all prime for an infinitude of natural numbers n, provided $\boldsymbol{h}$ is not a complete set of residues modulo any prime p; a naive interpretation of Cramér's model would have it that the same statement holds, without the proviso on $\boldsymbol{h}$. This discordance notwithstanding, Hardy–Littlewood and Cramér often agree: one could say that irregularities in the distribution of primes cancel each other out, over intervals on the right scale. An illustration of this trope lies in a conjectural answer to the following question. The prime number theorem asserts that, *on average* over natural numbers $n \leqslant x$, the number of primes in the interval $(n, n+\log x]$ is close to one, but for what *proportion* of $n \leqslant x$ does the interval contain the expected number of primes?

T. Freiberg (✉)
Department of Pure Mathematics, University of Waterloo, Waterloo, ON, Canada
e-mail: tfreiberg@uwaterloo.ca

© Springer International Publishing AG, part of Springer Nature 2018   23
J. Pintz, M. Th. Rassias (eds.), *Irregularities in the Distribution of Prime Numbers*,
https://doi.org/10.1007/978-3-319-92777-0_2

In Cramér's model, the sequence $(\mathbf{1}_{\mathbb{P}}(n))_{n \leqslant x}$, where $\mathbf{1}_{\mathbb{P}}$ is the characteristic function of the set of all primes, $\mathbb{P}$, behaves like a sequence $(\xi_n)_{n \leqslant x}$ of coin flips, whereby $\xi_n = 1$ with probability $1/\log x$. The proportion in question is thus the probability that $\xi_1 + \cdots + \xi_{\lfloor \log x \rfloor} = 1$, viz. $\frac{\lfloor \log x \rfloor}{\log x} \left(1 - \frac{1}{\log x}\right)^{\lfloor \log x \rfloor - 1} \sim e^{-1} \ (x \to \infty)$. An extension of this heuristic leads to the following conjecture.

Conjecture 1.1 Fix a positive integer r, then fix any positive numbers $\lambda_1, \ldots, \lambda_r$, and any nonnegative integers $m_1, \ldots, m_r$. As $x \to \infty$,

$$\frac{1}{x} \sum_{\substack{n \leqslant x \\ \#\mathbb{P} \cap (n, n + (\lambda_1 + \cdots + \lambda_j) \log x] = m_1 + \cdots + m_j \\ j = 1, \ldots, r}} 1 \sim \prod_{j=1}^{r} \left(e^{-\lambda_j} \frac{\lambda_j^{m_j}}{m_j!}\right).$$

In Conjecture 1.1, we may also put $\mathbf{1}_{\mathbb{P}}(n)$ instead of 1 in the summand, and normalize by $\pi(x) := \sum_{n \leqslant x} \mathbf{1}_{\mathbb{P}}(n)$ instead of x. (See Theorem 5.1 (b).)

Similarly, Cramér's model allows us to make a prediction concerning the distribution of gaps between consecutive primes.

Conjecture 1.2 Let $p_1 = 2 < p_2 = 3 < \cdots$ be the sequence of all primes. Fix a positive integer r, then fix any positive numbers $\lambda_1, \ldots, \lambda_r$. As $x \to \infty$,

$$\frac{1}{\pi(x)} \sum_{\substack{p_n \leqslant x \\ p_{n+j} - p_{n+j-1} \leqslant \lambda_j \log x \\ j = 1, \ldots, r}} 1 \sim \prod_{j=1}^{r} \left(1 - e^{-\lambda_j}\right).$$

For a detailed and insightful discussion of Cramér's model, and its ramifications for the distribution of primes in intervals, we refer the reader to Soundararajan's expository article [24].

Via circle method heuristics, Hardy and Littlewood [8, Conjecture B, p. 42] conjectured that the proportion of natural numbers $n \leqslant x$ for which $n + h_1, \ldots, n + h_k$ are all prime, where $h_1 < \cdots < h_k$ are distinct integers, is close to $\mathfrak{S}_h / (\log x)^k$, where $\mathfrak{S}_h$ is a certain quantity depending on $\boldsymbol{h} = \{h_1, \ldots, h_k\}$, which is positive unless $\boldsymbol{h}$ is a complete set of residues modulo any prime p. They show that the "singular series" $\mathfrak{S}_h$ can be written as a product over primes p (we take this as our definition):

$$\mathfrak{S}_h := \prod_p \left(\frac{p - \nu_h(p)}{p}\right)\left(\frac{p}{p-1}\right)^k, \tag{1.1}$$

where $\nu_h(p) := \#\{h + p\mathbb{Z} : h \in \boldsymbol{h}\}$ is the number of distinct congruence classes modulo p represented by $\boldsymbol{h}$. We note that, as $\nu_h(p) = k$ for all but the finitely many primes p, it follows that $\mathfrak{S}_h$ converges to a positive number, provided $\boldsymbol{h}$ is *admissible*, i.e. $\nu_h(p) < p$ for all primes p; otherwise, $\mathfrak{S}_h = 0$.

Conjecture 1.3 (Uniform Hardy–Littlewood Prime Tuples Conjecture) Fix any positive integer k, and any positive number λ. Let x be a parameter tending to infinity, and let $\boldsymbol{h} = \{h_1, \ldots, h_k\}$ be a set of integers with $0 \leqslant h_1 < \cdots < h_k \leqslant \lambda \log x$. If $\boldsymbol{h}$ is admissible, then, as $x \to \infty$,

$$\sum_{n \leqslant x} \mathbf{1}_{\mathbb{P}}(n + h_1) \cdots \mathbf{1}_{\mathbb{P}}(n + h_k) \sim \mathfrak{S}_{\boldsymbol{h}} \int_2^x \frac{dt}{(\log t)^k}. \tag{1.2}$$

As already mentioned, this is consistent with Cramér's model as far as the distribution of primes in intervals of average length is concerned.

Theorem 1.4 *If, for every fixed k and λ, the statement of Conjecture 1.3 holds true, then Conjectures 1.1 and 1.2 also hold true.*

This is essentially because, although Cramér's model basically says that $\mathfrak{S}_{\boldsymbol{h}}$ should always be equal to one, which is false, $\mathfrak{S}_{\boldsymbol{h}}$ is of *average order* one (see Theorem 4.4 below). In fact, as one may expect, an average version of Conjecture 1.3 is enough for Conjectures 1.1 and 1.2: see Theorems 5.1 and 5.2 below.

Theorem 1.4 is almost folklore: that a version of Conjecture 1.3 implies Conjecture 1.2 is mentioned in a 1972 survey article [11, p. 137] of Hooley; that a version of Conjecture 1.3 implies Conjecture 1.1 is established in a 1976 paper of Gallagher [5, Theorem 1]. Actually, in both [5] and [11], only the case $r = 1$ is mentioned explicitly, but it is clear that the arguments apply to the general case. Our main purpose is to present an alternative proof of Gallagher's result, which uses a direct inclusion–exclusion argument instead of a moment calculation. The advantage here is that one could easily track explicit error terms (conjectured and otherwise) through the entire argument, which may lead to more precise conjectures than Conjectures 1.1 and 1.2. (See Sect. 6.)

2 Some Literature

2.1 The Singular Series on Average

As alluded to above, the heart of the proof of a result like Theorem 1.4 is that the singular series, $\mathfrak{S}_{\boldsymbol{h}}$, is of average order one over $\boldsymbol{h}$ whose elements run over an interval, or certain geometric regions. There are many proofs of this fact in the literature: Gallagher [5] uses combinatorial identities for Stirling numbers of the second kind; Ford [3] and Pintz [22] simplify Gallagher's proof, leading to a result with a weaker error term; Soundararajan [24] uses contour integration to give a very precise estimate for the average of $\mathfrak{S}_{\boldsymbol{h}}$ in the case $\#\boldsymbol{h} = 2$; Montgomery and Soundararajan [21] (see Remark 4.8) extend this precise estimate using a fundamental inequality of Montgomery and Vaughan; Kowalski [13] develops a probabilistic framework for evaluating averages of singular series; David et al. [2]

develop a general method for estimating weighted sums involving singular series. However, our secondary purpose is to promulgate an elegant and flexible argument developed by Kurlberg et al. [7, 14–16].

2.2 Unconditional Results

Obviously Conjectures 1.1 and 1.2 are deep. As far as unconditional results in their direction, only a few weak approximations exist, and these have come only after significant advances in prime number theory.

Conjecture 1.2 asserts, in a quantitative way, that every nonnegative real number is a limit point of the sequence $(d_n/\log p_n)_{n \in \mathbb{N}}$, where $d_n := p_{n+1} - p_n$. Using Maier's matrix method, Hildebrand and Maier [9] show that a positive (but unspecified) proportion of nonnegative real numbers are limit points of this sequence. Using Maynard's celebrated work [19] on prime gaps, Banks et al. [1] show that the proportion is at least 12.5%, and Pintz [23] improves this to 25%. In spite of this progress, no limit point of $(d_n/\log p_n)_{n \in \mathbb{N}}$ is known, other than ∞ (this is an old result of Westzynthius [25]), and 0 (this is due to the breakthrough work of Goldston et al. [6]).

The result of Hildebrand and Maier extends to chains of gaps

$$(d_n/\log p_n, \ldots, d_{n+r-1}/\log p_n)_{n \in \mathbb{N}}.$$

They show that a positive (unspecified) proportion of tuples in $(0, \infty)^r$ are limit points. The work of Banks, Freiberg, and Maynard shows that there exist tuples in $(0, \infty)^r$ of a certain form that are limit points, but does not show that the proportion is positive when $r > 1$. It ought to be possible to obtain an explicit result here, of the same quality as the aforementioned result for the special case $r = 1$.

As for Conjecture 1.1, using Maynard's work [20], Freiberg [4] shows that the number of $n \leqslant x$ for which $(n, n + \lambda \log x]$ contains exactly m primes, for any given $\lambda > 0$ and integer $m \geqslant 0$, is at least $x^{1-o(1)}$ as $x \to \infty$. The proof is ad hoc: it produces intervals a bit longer than $\lambda \log x$, say $(n, n + 3\lambda \log x]$, which contain between m and $e^{O(m)}$ primes, and allows one to conclude that some subinterval of length $\lambda \log x$ has exactly m primes. This doesn't prima facie extend to a corresponding approximation to Conjecture 1.1 for $r > 1$, but again, such a result ought to be within the state of the art.

3 Notation

We use $\mathbb{R}_+$ to denote the set of positive real numbers. We define the natural numbers as $\mathbb{N} := \{1, 2, \ldots\}$, and denote the set of nonnegative integers by $\mathbb{N}_0$ (i.e., $\mathbb{N} \subsetneq \mathbb{N}_0 := \{0\} \cup \mathbb{N}$). The letter p stands for a prime, and p_n denotes the

nth smallest prime in the set of all primes, $\mathbb{P}$. Given $\mathscr{A} \subseteq \mathbb{Z}$, $\mathbf{1}_{\mathscr{A}} : \mathbb{Z} \to \{0, 1\}$ denotes the characteristic function of $\mathscr{A}$. Given $n \in \mathbb{N}$, $\omega(n)$ denotes the number of distinct prime divisors of n. We use $\sum^{\flat}$ to denote summation restricted to squarefree integers.

We view x and y as a real parameters tending to infinity. By $o(1)$ we mean a quantity that tends to zero as x (or y) tends to infinity; $A = (1 + o(1))B$ is also denoted by $A \sim B$; by $o(x)$ (respectively, $o(x/\log x)$) we mean a quantity A such that $A/x = o(1)$ (respectively, $A/(x/\log x) = o(1)$). Expressions of the form $A = O(B)$ and $A \ll B$ denote that $|A| \leqslant c|B|$, where c is some positive constant, throughout the domain of the quantity A. The constant c is to be regarded as absolute, and independent of any parameter unless indicated otherwise by subscripts, e.g. as in $A \ll_{k,\lambda} B$ (c depends on k and λ only). When we write $O_k(1)$ in an exponent, we mean a sufficiently large positive quantity, which depends on k.

4 Ancillary Results

4.1 Inclusion–Exclusion

The key to establishing Theorem 1.4, and Theorems 5.1 and 5.2 below, is showing that the singular series (1.1) is one on average (see Theorem 4.4). This fact is combined with Conjecture 1.3, or the weaker hypotheses elaborated in Sect. 5, via inclusion–exclusion.

Proposition 4.1 *Let $\mathscr{A} \subseteq \mathbb{Z}$. Let $\mathscr{B}$ be any finite set of integers, let $k_1, \ldots, k_r \in \mathbb{N}_0$, and let $y_1, \ldots, y_r \geqslant 1$. For all $m \in \mathbb{N}_0$, we have, with $y_0 := 0$,*

$$
\sum_{\substack{b \in \mathscr{B} \\ \#\mathscr{A} \cap (b + y_{i-1}, b + y_i] = k_i \\ i=1,\ldots,r}} 1 \geqslant \sum_{\substack{(\ell_1,\ldots,\ell_r) \in \mathbb{N}_0^r \\ \ell_1+\cdots+\ell_r \leqslant 2m+1}} \prod_{i=1}^{r} (-1)^{\ell_i} \binom{k_i + \ell_i}{\ell_i} \sum_{\substack{y_{i-1} < h_1^{(i)} < \cdots < h_{k_i+\ell_i}^{(i)} \leqslant y_i \\ i=1,\ldots,r}}
$$

$$
\times \sum_{b \in \mathscr{B}} \prod_{i=1}^{r} \mathbf{1}_{\mathscr{A}}\left(b + h_1^{(i)}\right) \cdots \mathbf{1}_{\mathscr{A}}\left(b + h_{k_i+\ell_i}^{(i)}\right),
$$

and

$$
\sum_{\substack{b \in \mathscr{B} \\ \#\mathscr{A} \cap (b + y_{i-1}, b + y_i] = k_i \\ i=1,\ldots,r}} 1 \leqslant \sum_{\substack{(\ell_1,\ldots,\ell_r) \in \mathbb{N}_0^r \\ \ell_1+\cdots+\ell_r \leqslant 2m}} \prod_{i=1}^{r} (-1)^{\ell_i} \binom{k_i + \ell_i}{\ell_i} \sum_{\substack{y_{i-1} < h_1^{(i)} < \cdots < h_{k_i+\ell_i}^{(i)} \leqslant y_i \\ i=1,\ldots,r}}
$$

$$
\times \sum_{b \in \mathscr{B}} \prod_{i=1}^{r} \mathbf{1}_{\mathscr{A}}\left(b + h_1^{(i)}\right) \cdots \mathbf{1}_{\mathscr{A}}\left(b + h_{k_i+\ell_i}^{(i)}\right).
$$

(In case $k_i + \ell_i = 0$ for some i, the condition $y_{i-1} < h_1^{(i)} < \cdots < h_{k_i+\ell_i}^{(i)} \leqslant y_i$ holds vacuously, and $\mathbf{1}_{\mathscr{A}}(b + h_1^{(i)}) \cdots \mathbf{1}_{\mathscr{A}}(b + h_{k_i+\ell_i}^{(i)})$ is an empty product [equal to one].)

Proof Evidently, for $(\ell_1, \ldots, \ell_r) \in \mathbb{N}_0^r$,

$$
\sum_{\substack{y_{i-1} < h_1^{(i)} < \cdots < h_{k_i+\ell_i}^{(i)} \leqslant y_i \\ i=1,\ldots,r}} \sum_{b \in \mathscr{B}} \prod_{i=1}^r \mathbf{1}_{\mathscr{A}}\big(b + h_1^{(i)}\big) \cdots \mathbf{1}_{\mathscr{A}}\big(b + h_{k_i+\ell_i}^{(i)}\big)
$$

$$
= \sum_{(j_1,\ldots,j_r) \in \mathbb{N}_0^r} \prod_{i=1}^r \binom{j_i + k_i + \ell_i}{k_i + \ell_i} \sum_{\substack{b \in \mathscr{B} \\ \#\mathscr{A} \cap (b+y_{i-1}, b+y_i] = j_i + k_i + \ell_i \\ i=1,\ldots,r}} 1,
$$

and hence, for every $M \in \mathbb{N}_0$,

$$
\sum_{\substack{(\ell_1,\ldots,\ell_r) \in \mathbb{N}_0^r \\ \ell_1+\cdots+\ell_r \leqslant M}} \prod_{i=1}^r (-1)^{\ell_i} \binom{k_i + \ell_i}{\ell_i} \sum_{\substack{y_{i-1} < h_1^{(i)} < \cdots < h_{k_i+\ell_i}^{(i)} \leqslant y_i \\ i=1,\ldots,r}} \sum_{b \in \mathscr{B}} \prod_{i=1}^r \mathbf{1}_{\mathscr{A}}\big(b + h_1^{(i)}\big) \cdots \mathbf{1}_{\mathscr{A}}\big(b + h_{k_i+\ell_i}^{(i)}\big)
$$

$$
= \sum_{\substack{(\ell_1,\ldots,\ell_r) \in \mathbb{N}_0^r \\ \ell_1+\cdots+\ell_r \leqslant M}} \prod_{i=1}^r (-1)^{\ell_i} \binom{k_i + \ell_i}{\ell_i} \sum_{(j_1,\ldots,j_r) \in \mathbb{N}_0^r} \prod_{i=1}^r \binom{j_i + k_i + \ell_i}{k_i + \ell_i}
$$

$$
\times \sum_{\substack{b \in \mathscr{B} \\ \#\mathscr{A} \cap (b+y_{i-1}, b+y_i] = j_i + k_i + \ell_i \\ i=1,\ldots,r}} 1
$$

$$
= \sum_{\substack{(\ell_1,\ldots,\ell_r) \in \mathbb{N}_0^r \\ \ell_1+\cdots+\ell_r \leqslant M}} \sum_{(j_1,\ldots,j_r) \in \mathbb{N}_0^r} \prod_{i=1}^r (-1)^{\ell_i} \binom{k_i + \ell_i}{\ell_i} \binom{j_i + k_i + \ell_i}{k_i + \ell_i} \sum_{\substack{b \in \mathscr{B} \\ \#\mathscr{A} \cap (b+y_{i-1}, b+y_i] = j_i + k_i + \ell_i \\ i=1,\ldots,r}} 1
$$

$$
= \sum_{\substack{(\ell_1,\ldots,\ell_r) \in \mathbb{N}_0^r \\ \ell_1+\cdots+\ell_r \leqslant M}} \sum_{(j_1,\ldots,j_r) \in \mathbb{N}_0^r} \prod_{i=1}^r (-1)^{\ell_i} \binom{j_i + k_i + \ell_i}{j_i + \ell_i} \binom{j_i + \ell_i}{\ell_i} \sum_{\substack{b \in \mathscr{B} \\ \#\mathscr{A} \cap (b+y_{i-1}, b+y_i] = j_i + k_i + \ell_i \\ i=1,\ldots,r}} 1
$$

$$
= \sum_{(n_1,\ldots,n_r) \in \mathbb{N}_0^r} \left\{ \left(\sum_{\substack{(\ell_1,\ldots,\ell_r) \in \mathbb{N}_0^r \\ \ell_1+\cdots+\ell_r \leqslant M}} \prod_{i=1}^r (-1)^{\ell_i} \binom{n_i}{\ell_i} \right) \right.
$$

$$
\left. \cdot \prod_{i=1}^r \binom{k_i + n_i}{n_i} \cdot \left(\sum_{\substack{b \in \mathscr{B} \\ \#\mathscr{A} \cap (b+y_{i-1}, b+y_i] = k_i + n_i \\ i=1,\ldots,r}} 1 \right) \right\}.
$$

Thus, for every $M \in \mathbb{N}_0$,

$$\sum_{\substack{b \in \mathscr{B} \\ \#\mathscr{A} \cap (b + y_{i-1}, b + y_i] = k_i \\ i=1,\dots,r}} 1$$

$$= \sum_{\substack{(\ell_1,\dots,\ell_r) \in \mathbb{N}_0^r \\ \ell_1 + \dots + \ell_r \leqslant M}} \prod_{i=1}^{r} (-1)^{\ell_i} \binom{k_i + \ell_i}{\ell_i} \sum_{\substack{y_{i-1} < h_1^{(i)} < \dots < h_{k_i+\ell_i}^{(i)} \leqslant y_i \\ i=1,\dots,r}} \sum_{b \in \mathscr{B}} \prod_{i=1}^{r} \mathbf{1}_{\mathscr{A}}(b + h_1^{(i)}) \cdots \mathbf{1}_{\mathscr{A}}(b + h_{k_i+\ell_i}^{(i)})$$

$$- \sum_{\substack{(n_1,\dots,n_r) \in \mathbb{N}_0^r \\ n_1 + \dots + n_r \geqslant 1}} \left\{ \left(\sum_{\substack{(\ell_1,\dots,\ell_r) \in \mathbb{N}_0^r \\ \ell_1 + \dots + \ell_r \leqslant M}} \prod_{i=1}^{r} (-1)^{\ell_i} \binom{n_i}{\ell_i} \right) \cdot \binom{k_i + n_i}{n_i} \cdot \left(\sum_{\substack{b \in \mathscr{B} \\ \#\mathscr{A} \cap (b + y_{i-1}, b + y_i] = k_i + n_i \\ i=1,\dots,r}} 1 \right) \right\}.$$

The result follows in view of the claim, which we will presently verify, that for any $M, n_1, \dots, n_r \in \mathbb{N}_0$ with $n_1 + \dots + n_r \geqslant 1$,

$$(-1)^M \sum_{\substack{(\ell_1,\dots,\ell_r) \in \mathbb{N}_0^r \\ \ell_1 + \dots + \ell_r \leqslant M}} \prod_{i=1}^{r} (-1)^{\ell_i} \binom{n_i}{\ell_i} \geqslant 0. \tag{4.1}$$

To verify this claim, we induct on r. For $n, N \in \mathbb{N}_0$, we have the standard identity

$$\sum_{\ell=0}^{N} (-1)^{\ell} \binom{n}{\ell} = (-1)^N \binom{n-1}{N}, \tag{4.2}$$

which gives (4.1) in the special case $r = 1$. Assume the claim holds for an arbitrary r. Writing L for $\ell_1 + \dots + \ell_r$, we have, in view of (4.2),

$$\sum_{\substack{(\ell_1,\dots,\ell_{r+1}) \in \mathbb{N}_0^{r+1} \\ \ell_1 + \dots + \ell_{r+1} \leqslant M}} \prod_{i=1}^{r+1} (-1)^{\ell_i} \binom{n_i}{\ell_i} = \sum_{\substack{(\ell_1,\dots,\ell_r) \in \mathbb{N}_0^r \\ \ell_1 + \dots + \ell_r \leqslant M}} \prod_{i=1}^{r} (-1)^{\ell_i} \binom{n_i}{\ell_i} \sum_{\ell_{r+1}=0}^{M-L} (-1)^{\ell_{r+1}} \binom{n_{r+1}}{M-L}$$

$$= (-1)^M \sum_{\substack{(\ell_1,\dots,\ell_r) \in \mathbb{N}_0^r \\ \ell_1 + \dots + \ell_r \leqslant M}} \prod_{i=1}^{r} \binom{n_i}{\ell_i} \binom{n_{r+1}-1}{M-L}.$$

If $n_{r+1} \geqslant 1$, then each term in this last summand is nonnegative, and hence (4.1) holds with $r + 1$ in place of r. If $n_{r+1} = 0$, then $n_1 + \dots + n_{r+1} \geqslant 1$ is equivalent to $n_1 + \dots + n_r \geqslant 1$, and

$$\sum_{\substack{(\ell_1,\ldots,\ell_r) \in \mathbb{N}_0^r \\ \ell_1+\cdots+\ell_r \leqslant M}} \prod_{i=1}^{r} \binom{n_i}{\ell_i}\binom{n_{r+1}-1}{M-L} = \sum_{\substack{(\ell_1,\ldots,\ell_r) \in \mathbb{N}_0^r \\ \ell_1+\cdots+\ell_r \leqslant M}} \prod_{i=1}^{r}(-1)^{\ell_i}\binom{n_i}{\ell_i},$$

in which case (4.1) holds with $r+1$ in place of r, by inductive hypothesis. $\qquad\square$

The following inclusion–exclusion argument is given by Hooley in [10], for $r = 1$ (the case $r \geqslant 2$ is implicit), in a more abstract form by Kurlberg and Rudnick in [16, Appendix A], and in a highbrow form by Katz and Sarnak in [12, Key Lemma 2.4.12].

Proposition 4.2 *Let $\mathscr{A} \subseteq \mathbb{Z}$ be bounded from below, and let $a_1 < a_2 < \cdots$ be the sequence of all integers in $\mathscr{A}$ in ascending order. Let $\mathscr{B}$ be a finite subset of $\mathscr{A}$. Let $r \in \mathbb{N}$, and let $y_1,\ldots,y_r \geqslant 1$. For all $m \in \mathbb{N}_0$, we have*

$$\sum_{\substack{a_t \in \mathscr{B} \\ a_{t+j}-a_{t+j-1} \leqslant y_j \\ j=1,\ldots,r}} 1 \geqslant \sum_{k=r}^{r+2m+1} (-1)^{k-r} \sum_{\substack{(k_1,\ldots,k_r) \in \mathbb{N}^r \\ k_1+\cdots+k_r=k}} \sum_{\substack{0<h_1<\cdots<h_k \\ h_{k_1+\cdots+k_j}-h_{k_1+\cdots+k_{j-1}} \leqslant y_j \\ j=1,\ldots,r}}$$

$$\times \sum_{b \in \mathscr{B}} \mathbf{1}_{\mathscr{A}}(b+h_1)\cdots\mathbf{1}_{\mathscr{A}}(b+h_k),$$

and

$$\sum_{\substack{a_t \in \mathscr{B} \\ a_{t+j}-a_{t+j-1} \leqslant y_j \\ j=1,\ldots,r}} 1 \leqslant \sum_{k=r}^{r+2m} (-1)^{k-r} \sum_{\substack{(k_1,\ldots,k_r) \in \mathbb{N}^r \\ k_1+\cdots+k_r=k}} \sum_{\substack{0<h_1<\cdots<h_k \\ h_{k_1+\cdots+k_j}-h_{k_1+\cdots+k_{j-1}} \leqslant y_j \\ j=1,\ldots,r}}$$

$$\times \sum_{b \in \mathscr{B}} \mathbf{1}_{\mathscr{A}}(b+h_1)\cdots\mathbf{1}_{\mathscr{A}}(b+h_k).$$

Proof Given $\hat{\boldsymbol{k}} = (k_1,\ldots,k_r)$ and $\hat{\boldsymbol{c}} = (c_1,\ldots,c_r)$ in $\mathbb{N}^r$, let

$$N_{\hat{\boldsymbol{k}},\hat{\boldsymbol{c}}}(\mathscr{B}) := \sum_{\substack{0<h_1<\cdots<h_{k_1+\cdots+k_r} \\ h_{k_1+\cdots+k_j}=c_j \\ j=1,\ldots,r}} \sum_{b \in \mathscr{B}} \mathbf{1}_{\mathscr{A}}(b+h_1)\cdots\mathbf{1}_{\mathscr{A}}(b+h_{k_1+\cdots+k_r}).$$

Let m be any nonnegative integer. We claim (and will verify) that, for any given $\hat{\boldsymbol{c}} = (c_1,\ldots,c_r) \in \mathbb{N}^r$ with $c_1 < \cdots < c_r$,

$$\sum_{k=r}^{r+2m+1} (-1)^{k-r} \sum_{\substack{(k_1,\ldots,k_r) \in \mathbb{N}^r \\ k_1+\cdots+k_r=k}} N_{\hat{k},\hat{c}}(\mathscr{B})$$

$$\leqslant \sum_{\substack{a_t \in \mathscr{B} \\ a_{t+j}-a_{t+j-1}=c_j \\ j=1,\ldots,r}} 1 \quad \leqslant \quad \sum_{k=r}^{r+2m} (-1)^{k-r} \sum_{\substack{(k_1,\ldots,k_r) \in \mathbb{N}^r \\ k_1+\cdots+k_r=k}} N_{\hat{k},\hat{c}}(\mathscr{B}). \tag{4.3}$$

To deduce the result from our claim, we note that

$$\sum_{\substack{a_t \in \mathscr{B} \\ a_{t+j}-a_{t+j-1} \leqslant y_j \\ j=1,\ldots,r}} 1 = \sum_{\substack{\hat{c} \in \mathbb{N}^r \\ 0<c_j-c_{j-1} \leqslant y_j \\ j=1,\ldots,r}} \sum_{\substack{a_t \in \mathscr{B} \\ a_{t+j}-a_{t+j-1}=c_j \\ j=1,\ldots,r}} 1, \tag{4.4}$$

wherein $\hat{c} = (c_1,\ldots,c_r)$, and $c_0 := 0$. Given $\hat{k} \in \mathbb{N}^r$ with $k_1 + \cdots + k_r = k$,

$$\sum_{\substack{\hat{c} \in \mathbb{N}^r \\ 0<c_j-c_{j-1} \leqslant y_j \\ j=1,\ldots,r}} N_{\hat{k},\hat{c}}(\mathscr{B}) = \sum_{\substack{0<h_1<\cdots<h_k \\ h_{k_1+\cdots+k_j}-h_{k_1+\cdots+k_{j-1}} \leqslant y_j \\ j=1,\ldots,r}} \sum_{b \in \mathscr{B}} \mathbf{1}_{\mathscr{A}}(b+h_1)\cdots\mathbf{1}_{\mathscr{A}}(b+h_k). \tag{4.5}$$

Combining (4.3), (4.4), and (4.5), then changing order of summation, we obtain the inequalities in the statement of the proposition.

We now prove our claim (4.3). Let $\hat{c} \in \mathbb{N}^r$ with $c_1 < \cdots < c_r$ be given. First of all note that, for any $\hat{k} \in \mathbb{N}^r$,

$$N_{\hat{k},\hat{c}}(\mathscr{B}) = \sum_{b \in \mathscr{B}} \sum_{\substack{0<h_1<\cdots<h_{k_1+\cdots+k_r} \\ h_{k_1+\cdots+k_j}=c_j \\ j=1,\ldots,r}} \mathbf{1}_{\mathscr{A}}(b+h_1)\cdots\mathbf{1}_{\mathscr{A}}(b+h_{k_1+\cdots+k_r}). \tag{4.6}$$

For $b \in \mathbb{Z}$, let $M_j(b,c_j)$ be the number of elements of $\mathscr{A}$ in-between $b+c_{j-1}$ and $b+c_j$, i.e.

$$M_j(b) = M_j(b,c_j) := \sum_{b+c_{j-1}<a<b+c_j} \mathbf{1}_{\mathscr{A}}(a),$$

for $j = 1,\ldots,r$, where $c_0 := 0$. From (4.6) it is not difficult to see that, for any $\hat{k} \in \mathbb{N}^r$,

$$N_{\hat{k},\hat{c}}(\mathscr{B}) = \sum_{b \in \mathscr{B}} \left\{ \mathbf{1}_{\mathscr{A}}(b+c_1)\cdots\mathbf{1}_{\mathscr{A}}(b+c_r) \binom{M_1(b)}{k_1-1}\cdots\binom{M_r(b)}{k_r-1} \right\}. \tag{4.7}$$

Given $b \in \mathscr{B}$ and $b + c_1, \ldots, b + c_r \in \mathscr{A}$, having $M_1(b) = \cdots = M_r(b) = 0$ is equivalent to $b, b + c_1, \ldots, b + c_r$ being *consecutive* elements of $\mathscr{A}$, i.e. for some t, $b = a_t$ and $a_{t+j} - a_{t+j-1} = c_j$, $j = 1, \ldots, r$.

Next, for any integer $k \geqslant r$ and any nonnegative integers $M_1, \ldots, M_r$, we have

$$\sum_{\substack{(k_1,\ldots,k_r) \in \mathbb{N}^r \\ k_1 + \cdots + k_r = k}} \binom{M_1}{k_1 - 1} \cdots \binom{M_r}{k_r - 1} = \binom{M_1 + \cdots + M_r}{k - r}. \tag{4.8}$$

To see this, note that each summand on the left-hand side is the number of ways of choosing $k - r$ objects from a set X of size $M_1 + \cdots + M_r$, in such a way that $k_j - 1$ objects are chosen from a subset $X_j \subseteq X$ of size M_j, and where $X = X_1 \cup \cdots \cup X_r$ is a partition. Summing over all partitions of k into r positive integers, we end up with the total number of ways to choose $k - r$ objects from X, viz. the right-hand side. Also,

$$\sum_{k-r \geqslant 0} (-1)^{k-r} \binom{M_1 + \cdots + M_r}{k - r} = \begin{cases} 1 & M_1 + \cdots + M_r = 0, \\ 0 & \text{otherwise,} \end{cases} \tag{4.9}$$

the left-hand side being the binomial expansion of $(1 - 1)^{M_1 + \cdots + M_r}$ in the second case. Furthermore, for any nonnegative integers $M_1, \ldots, M_r$ we have

$$\sum_{k-r=0}^{2m+1} (-1)^{k-r} \binom{M_1 + \cdots + M_r}{k - r}$$

$$\leqslant \sum_{k-r \geqslant 0} (-1)^{k-r} \binom{M_1 + \cdots + M_r}{k - r} \leqslant \sum_{k-r=0}^{2m} (-1)^{k-r} \binom{M_1 + \cdots + M_r}{k - r}, \tag{4.10}$$

as can be verified by using the recurrence relation $\binom{M}{i} = \binom{M-1}{i} + \binom{M-1}{i-1}$.

Combining (4.7)–(4.9), we find, after changing order of summation, that

$$\sum_{\substack{b \in \mathscr{B} \\ \text{consecutive}}} \mathbf{1}_{\mathscr{A}}(b + c_1) \cdots \mathbf{1}_{\mathscr{A}}(b + c_r) = \sum_{k-r \geqslant 0} (-1)^{k-r} \sum_{\substack{(k_1,\ldots,k_r) \in \mathbb{N}^r \\ k_1 + \cdots + k_r = k}} N_{\hat{k}, \hat{c}}(\mathscr{B}), \tag{4.11}$$

where in the summand on the left-hand side, "consecutive" indicates summation restricted to those b for which, for some t, $b = a_t$ and $a_{t+j} - a_{t+j-1} = c_j$, $j = 1, \ldots, r$. Combining (4.7), (4.8), and (4.10), we similarly find that

$$\sum_{k-r=0}^{2m+1} (-1)^{k-r} \sum_{\substack{(k_1,\dots,k_r) \in \mathbb{N}^r \\ k_1+\cdots+k_r=k}} N_{\hat{k},\hat{c}}(\mathscr{B})$$

$$\leqslant \sum_{\substack{b \in \mathscr{B} \\ \text{consecutive}}} \mathbf{1}_{\mathscr{A}}(b+c_1)\cdots\mathbf{1}_{\mathscr{A}}(b+c_r) \leqslant \sum_{k-r=0}^{2m} (-1)^{k-r} \sum_{\substack{(k_1,\dots,k_r) \in \mathbb{N}^r \\ k_1+\cdots+k_r=k}} N_{\hat{k},\hat{c}}(\mathscr{B}).$$

$$(4.12)$$

These are the claimed inequalities in (4.3), since

$$\sum_{\substack{b \in \mathscr{B} \\ \text{consecutive}}} \mathbf{1}_{\mathscr{A}}(b+c_1)\cdots\mathbf{1}_{\mathscr{A}}(b+c_r) = \sum_{\substack{a_t \in \mathscr{B} \\ a_{t+j}-a_{t+j-1}=c_j \\ j=1,\dots,r}} 1.$$

$$\square$$

4.2   The Singular Series, As a Series

Given $h \subseteq \mathbb{Z}$ and $j \in \mathbb{N}_0$, define a multiplicative function $d \mapsto \epsilon_h(d; j)$ by letting, for $p \in \mathbb{P}$ and $d \in \mathbb{N}$,

$$\epsilon_h(p; j) := \left(\frac{p - \nu_h(p)}{p}\right)\left(\frac{p}{p-1}\right)^j - 1 \quad \text{and} \quad \epsilon_h(d; j) := \prod_{p|d} \epsilon_h(p; j),$$

$$(4.13)$$

with the convention that $\epsilon_h(1; j) := 1$. Here, as in (1.1),

$$\nu_h(p) := \#\{h + p\mathbb{Z} : h \in h\}. \tag{4.14}$$

If h is finite with $k := \#h$, then (see (1.1))

$$\mathfrak{S}_h := \prod_p (1 + \epsilon_h(p)) = \sum_{d \geqslant 1}^{\flat} \epsilon_h(d; k), \tag{4.15}$$

the sum being absolutely convergent in view of the next proposition. In this proposition and below,

$$\det(h) := \prod_{\substack{h, h' \in h \\ h < h'}} (h' - h). \tag{4.16}$$

(We have $\nu_h(p) = k$ when $p \nmid \det(h)$.) Also recall that $\sum^{\flat}$ denotes summation restricted to squarefrees, and $\omega(d) := \sum_{p|d} 1$.

Proposition 4.3

(a) *Let $\boldsymbol{h} \subset \mathbb{Z}$ be finite, and let $k := \#\boldsymbol{h}$. There exists $A_k \in \mathbb{N}$, depending on k, such that for all squarefree $d \in \mathbb{N}$,*

$$|\epsilon_{\boldsymbol{h}}(d; k)| \leqslant A_k^{\omega(d)} \frac{(d, \det(\boldsymbol{h}))}{d^2}.$$

(b) *Fix $A \in \mathbb{N}$. For $x \geqslant 3$,*

$$\sum_{d > x}^{\flat} \frac{A^{\omega(d)}}{d^2} \ll_A \frac{(\log x)^{A-1}}{x}.$$

Proof This is a routine verification, so let us spare the reader the details.   □

4.3   The Singular Series, on Average

Given $k \in \mathbb{N}$, let

$$\Delta^k := \{(x_1, \ldots, x_k) \in \mathbb{R}^k : 0 < x_1 < \cdots < x_k\}; \tag{4.17}$$

given $\mathscr{C} \subseteq \mathbb{R}^k$ and $y \in \mathbb{R}$, let

$$y\mathscr{C} := \{(yx_1, \ldots, yx_k) : (x_1, \ldots, x_k) \in \mathscr{C}\}.$$

Theorem 4.4 *Set $\boldsymbol{o} := \varnothing$, or set $\boldsymbol{o} := \{0\}$. Fix $k \in \mathbb{N}$ and a bounded convex set $\mathscr{C} \subset \Delta^k$ (see (4.17)). For $y \geqslant 3$,*

$$\sum_{(h_1, \ldots, h_k) \in y\mathscr{C} \cap \mathbb{Z}^k} \mathfrak{S}_{\boldsymbol{o} \cup \boldsymbol{h}} = y^k \mathrm{vol}(\mathscr{C}) + O_{k,\mathscr{C}}\left(y^{k-1}(\log y)^{O_k(1)}\right), \tag{4.18}$$

where $\boldsymbol{h} = \{h_1, \ldots, h_k\}$ in the summand, and vol *stands for volume in $\mathbb{R}^k$.*

The proof of Theorem 4.4 presented here is essentially due to Kurlberg et al. [7, 14–16]. It is simple and quite generalizable, although it could be more precise (see Remark 4.8). The proof uses the next three lemmas: the crux of it is the cancellation in Lemma 4.7.

Lemma 4.5 *Fix $k \in \mathbb{N}$, then fix $A_k \in \mathbb{N}$, depending on k only. For $x \geqslant 3$,*

$$\sum_{d > x}^{\flat} \frac{A_k^{\omega(d)}}{d^2} \sum_{0 < h_1 < \cdots < h_k \leqslant x} (d, \det(\{0, h_1, \ldots, h_k\})) \ll_k x^{k-1}(\log x)^{O_k(1)}.$$

Proof Let $x \geqslant 3$. We first show that, for any squarefree $c \in \mathbb{N}$,

$$\sum_{\substack{0 < h_1 < \cdots < h_k \leqslant x \\ c \mid \det(\{0, h_1, \ldots, h_k\})}} 1 \leqslant k^{2\omega(c)} \left(\frac{x^k}{c} + O_k(x^{k-1}) \right). \tag{4.19}$$

Let $h_0 = 0, h_1, \ldots, h_k$ be pairwise distinct integers, and suppose that c divides $\prod_{0 \leqslant i < j \leqslant k}(h_i - h_j)$. Then, since c is squarefree, there exist pairwise coprime positive integers c_{ij} such that $c = \prod_{0 \leqslant i < j \leqslant k} c_{ij}$ and $c_{ij} \mid h_i - h_j, 0 \leqslant i < j \leqslant k$. Therefore,

$$\sum_{\substack{0 < h_1 < \cdots < h_k \leqslant x \\ c \mid \det(\{h_0, h_1, \ldots, h_k\})}} 1 \leqslant \sum_{c = c_{01} \cdots c_{(k-1)k}} \sum_{\substack{h_1 \in I_x \\ 0 \leqslant i < j \leqslant k-1 \Rightarrow c_{ij} \mid h_i - h_j}} \sum_{h_2 \in I_x} \cdots \sum_{h_{k-1} \in I_x} \sum_{\substack{h_k \in I_x \\ 0 \leqslant i \leqslant k-1 \Rightarrow c_{ik} \mid h_i - h_k}} 1,$$

where on the right-hand side, the outermost sum is over all decompositions of c as a product of $\binom{k+1}{2}$ positive integers, and $I_x := (0, x]$.

Consider the decomposition $c = c_{01} \cdots c_{(k-1)k}$. Let us define $c_j := \prod_{i=0}^{j-1} c_{ij}$ for $j = 1, \ldots, k$. Notice that $c = \prod_{j=1}^{k} c_j$. By the Chinese remainder theorem, the condition on h_k in the innermost sum above is equivalent to h_k being in some congruence class modulo c_k, uniquely determined by $h_0, h_1, \ldots, h_{k-1}$. The sum is therefore equal to $x/c_k + O(1)$. Iterating this argument k times, we see that the inner sum over $h_1, \ldots, h_k$ is equal to

$$\prod_{j=1}^{k} \left(\frac{x}{c_j} + O(1) \right) = \frac{x^k}{c} + O_k(x^{k-1}).$$

The bound (4.19) follows by combining and noting that, since c is squarefree, the number of ways of writing c as a product of $\binom{k+1}{2}$ positive integers is $\binom{k+1}{2}^{\omega(c)}$, and that $\binom{k+1}{2} \leqslant k^2$.

For pairwise distinct, nonzero integers $h_1, \ldots, h_k$, and any $d \in \mathbb{N}$, we have, trivially, $(d, \det(\{0, h_1, \ldots, h_k\})) \leqslant \sum_{c \mid d,\, c \mid \det(\{0, h_1, \ldots, h_k\})} c$. If $h_1, \ldots, h_k \leqslant x$ as well, then $p \mid c$ implies $p \leqslant x$ for c in this summand. From this and (4.19), it follows that

$$\sum_{0 < h_1 < \cdots < h_k \leqslant x} (d, \det\{0, h_1, \ldots, h_k\}) \ll_k x^k \sum_{c \mid d} k^{2\omega(c)} + x^{k-1} \sum_{\substack{c \mid d \\ p \mid c \Rightarrow p \leqslant x}} c k^{2\omega(c)}.$$

Now, for $c \mid d$ we have $k^{2\omega(c)} \leqslant k^{2\omega(d)}$, and, if d is squarefree, then $\sum_{c \mid d} 1 = 2^{\omega(d)}$. Applying these bounds to the left-hand side of the bound in the statement of the lemma, we see that it is

$$\ll_k x^k \sum_{d>x}^{\flat} \frac{B_k^{\omega(d)}}{d^2} + x^{k-1} \sum_{d\geqslant 1}^{\flat} \frac{B_k^{\omega(d)}}{d^2} \sum_{\substack{c\mid d \\ p\mid c \Rightarrow p\leqslant x}} c, \tag{4.20}$$

where $B_k \in \mathbb{N}$ is sufficiently large in terms of k (we can take $B_k = 2k^2 A_k$).

By Proposition 4.3 (b),

$$\sum_{d>x}^{\flat} \frac{B_k^{\omega(d)}}{d^2} \ll_k \frac{(\log x)^{B_k-1}}{x}. \tag{4.21}$$

Also,

$$\sum_{d\geqslant 1}^{\flat} \frac{B_k^{\omega(d)}}{d^2} \sum_{\substack{c\mid d \\ p\mid c \Rightarrow p\leqslant x}} c \leqslant \sum_{\substack{c\geqslant 1 \\ p\mid c \Rightarrow p\leqslant x}}^{\flat} \frac{B_k^{\omega(c)}}{c} \sum_{b\geqslant 1}^{\flat} \frac{B_k^{\omega(b)}}{b^2} \ll_k \sum_{\substack{c\geqslant 1 \\ p\mid c \Rightarrow p\leqslant x}}^{\flat} \frac{B_k^{\omega(c)}}{c};$$

as can be seen by writing $d = bc$ and changing order of summation; also

$$\sum_{\substack{c\geqslant 1 \\ p\mid c \Rightarrow p\leqslant x}}^{\flat} \frac{B_k^{\omega(c)}}{c} \leqslant \prod_{p\leqslant x}\left(1 + \frac{B_k}{p}\right) \leqslant \prod_{p\leqslant x}\left(1 + \frac{1}{p}\right)^{B_k} \ll_k (\log x)^{B_k}$$

by Mertens' theorem. Combining gives

$$\sum_{d\geqslant 1}^{\flat} \frac{B_k^{\omega(d)}}{d^2} \sum_{\substack{c\mid d \\ p\mid c \Rightarrow p\leqslant x}} c \ll_k (\log x)^{B_k}. \tag{4.22}$$

Finally, we obtain the result by combining (4.20) with (4.21) and (4.22). □

Lemma 4.6 *Fix $k \in \mathbb{N}$, and a bounded convex set $\mathscr{C} \subset \mathbb{R}^k$. For $y \geqslant 1$,*

$$\#(y\mathscr{C} \cap \mathbb{Z}^k) = y^k \mathrm{vol}(\mathscr{C}) + O_{k,\mathscr{C}}(y^{k-1}).$$

Proof See [17, pp. 128–129]. □

Lemma 4.7 *Let $f \subset \mathbb{Z}$ be finite, and let $l := \#f$. Let $R_1, \ldots, R_m$ be complete residue systems modulo a squarefree $d \in \mathbb{N}$. If $d \neq 1$, then*

$$\sum_{g_1\in R_1} \cdots \sum_{g_m\in R_m} \epsilon_{f\cup g}(d; l+m) = d^m \epsilon_f(d; l), \tag{4.23}$$

where $g = \{g_1, \ldots, g_m\}$ in the summand. In particular, if f is empty or a singleton, then the right-hand side of (4.23) is equal to zero.

Proof We may assume that $R_1 = \cdots = R_k = \{0, \ldots, d-1\}$; by multiplicativity (see (4.13)) and the Chinese remainder theorem, we have

$$\sum_{g_1 \in R_1} \cdots \sum_{g_m \in R_m} \epsilon_{f \cup g}(d; l+m) = \prod_{p \mid d} \left(\sum_{0 \leqslant g_1 < p} \cdots \sum_{0 \leqslant g_m < p} \epsilon_{f \cup g}(p; l+m) \right).$$

(Here and below, $g = \{g_1, \ldots, g_m\}$.) It therefore suffices to establish (4.23) for $d = p$, a prime. Now,

$$p - \nu_{f \cup g}(p) = \#\{n \bmod p : \forall h \in f \cup g, n + h \not\equiv 0 \bmod p\},$$

see (4.14); making this substitution and changing order of summation yields

$$\sum_{0 \leqslant g_1 < p} \cdots \sum_{0 \leqslant g_m < p} (p - \nu_{f \cup g}(p)) = \sum_{\substack{0 \leqslant n < p \\ \forall f \in f, n+f \not\equiv 0 \bmod p}} \sum_{\substack{0 \leqslant g_1 < p \\ n+g_1 \not\equiv 0 \bmod p}}$$

$$\cdots \sum_{\substack{0 \leqslant g_m < p \\ n+g_m \not\equiv 0 \bmod p}} 1,$$

which is equal to $(p - \nu_f(p))(p-1)^m$. This, and recalling the definition (4.13), (4.14), of $\epsilon_f(p; l+m)$ and $\epsilon_f(p; l)$, leads directly to the result. □

Proof of Theorem 4.4 Let $y \geqslant 3$. Let us write $\sum_h$ for $\sum_{(h_1,\ldots,h_k) \in y\mathscr{C} \cap \mathbb{Z}^k}$; h for $\{h_1, \ldots, h_k\}$; $\epsilon_{o \cup h}(d)$ for $\epsilon_{o \cup h}(d; \#o + k)$. Note that $0 < h_1 < \cdots < h_k \ll_{\mathscr{C}} y$ for all h with $(h_1, \ldots, h_k) \in y\mathscr{C} \cap \mathbb{Z}^k$. Let $A_{k,1}, A_{k,2}, A_{k,3}, A_{k,4}, A_{k,5}$ stand for numbers that are sufficiently large in terms of k.

By (4.15),

$$\sum_h \mathfrak{S}_{o \cup h} = \sum_h 1 + \sum_{1 < d \leqslant y}^{\flat} \sum_h \epsilon_{o \cup h}(d) + \sum_{d > y}^{\flat} \sum_h \epsilon_{o \cup h}(d), \tag{4.24}$$

and by Lemma 4.6,

$$\sum_h 1 = y^k \mathrm{vol}(\mathscr{C}) + O_{k,\mathscr{C}}(y^{k-1}). \tag{4.25}$$

By Proposition 4.3 (a) and Lemma 4.5,

$$\sum_{d > y}^{\flat} \sum_h |\epsilon_{o \cup h}(d)| \leqslant \sum_{d > y}^{\flat} \sum_h \frac{A_{k,1}^{\omega(d)}}{d^2}(d, \det(o \cup h)) \ll_{k,\mathscr{C}} y^{k-1} (\log y)^{A_{k,2}}. \tag{4.26}$$

Consider the middle sum on the right-hand side of (4.24). Let $d \in \mathbb{N}$ be squarefree with $d \leqslant y$, and partition $\mathbb{R}^k$ into cubes

$$C_{d,\hat{\imath}} := \{(x_1, \ldots, x_k) \in \mathbb{R}^k : t_i d \leqslant x_i < (t_i + 1)d, i = 1, \ldots, k\},$$

with $\hat{\imath} := (t_1, \ldots, t_k)$ running over $\mathbb{Z}^k$. Each $(h_1, \ldots, h_k)$ is a point in a unique cube of this form: we call $(h_1, \ldots, h_k)$ a d-*interior* point if this cube is entirely contained in $y\mathscr{C}$, and a d-*boundary* point otherwise. As $(h_1, \ldots, h_k)$ runs over all d-interior points of $y\mathscr{C}$, h_i, $i = 1, \ldots, k$, runs over a pairwise disjoint union of complete residue systems modulo d, none of which contain 0 (so $\#(o \cup h) = \#o + k$). By Lemma 4.7, it follows that

$$\sideset{}{^\mathrm{b}}\sum_{1<d\leqslant y} \sum_{h} \epsilon_{o \cup h}(d) = \sideset{}{^\mathrm{b}}\sum_{1<d\leqslant y} \sum_{\substack{h \\ d\text{-boundary}}} \epsilon_{o \cup h}(d). \tag{4.27}$$

By Proposition 4.3 (a), and since $(d, \det(o \cup h)) \leqslant \sum_{c|d,\, c|\det(\{0,h_1,\ldots,h_k\})} c$,

$$\sideset{}{^\mathrm{b}}\sum_{1<d\leqslant y} \sum_{\substack{h \\ d\text{-boundary}}} |\epsilon_{o \cup h}(d)| \leqslant \sideset{}{^\mathrm{b}}\sum_{d\leqslant y} \frac{A_{k,3}^{\omega(d)}}{d^2} \sum_{\substack{h \\ d\text{-boundary}}} (d, \det(o \cup h)) \leqslant \sideset{}{^\mathrm{b}}\sum_{d\leqslant y} \frac{A_{k,3}^{\omega(d)}}{d^2} \sum_{c|d} c$$

$$\times \sum_{\substack{h \\ d\text{-boundary} \\ c|\det(\{0,h_1,\ldots,h_k\})}} 1.$$

For each $d \in \mathbb{N}$ with $y/d \geqslant 1$, there are $\ll_{k,\mathscr{C}} (y/d)^{k-1}$ cubes $C_{d,\hat{\imath}}$ that have a nonempty intersection with the boundary of $y\mathscr{C}$. (See the proof of Lemma 4.6 in [17, pp. 128–129].) For each such boundary cube $C_{d,\hat{\imath}}$, the corresponding d-boundary points are all in $C_{d,\hat{\imath}} \cap \mathbb{Z}^k$, which is a product of complete residue systems modulo d, and, given that $c \mid d$, the condition $c \mid \det(\{0, h_1, \ldots, h_k\})$ is equivalent to $c \mid \det(\{0, h_1', \ldots, h_k'\})$ when $h_i \equiv h_i' \bmod d$, $i = 1, \ldots, k$.

Thus, for squarefree $d \leqslant y$, (4.19) yields ($c \mid d$ implies c is squarefree)

$$\sum_{\substack{h \\ d\text{-boundary} \\ c|\det(\{0,h_1,\ldots,h_k\})}} 1 \ll_{k,\mathscr{C}} \frac{y^{k-1}}{d^{k-1}} \sum_{\substack{0<h_1<\cdots<h_k\leqslant d \\ c|\det(\{0,h_1,\ldots,h_k\})}} 1 \ll_k y^{k-1} d\left(\frac{A_{k,4}^{\omega(c)}}{c}\right),$$

and hence

$$\sideset{}{^\mathrm{b}}\sum_{1<d\leqslant y} \sum_{\substack{h \\ d\text{-boundary}}} |\epsilon_{o \cup h}(d)| \ll_{k,\mathscr{C}} y^{k-1} \sideset{}{^\mathrm{b}}\sum_{d\leqslant y} \frac{A_{k,3}^{\omega(d)}}{d} \sum_{c|d} A_{k,4}^{\omega(c)} \leqslant y^{k-1} \sideset{}{^\mathrm{b}}\sum_{d\leqslant y} \frac{A_{k,5}^{\omega(d)}}{d},$$

since $\sum_{c|d} A_{k,4}^{\omega(c)}$ is at most $A_{k,4}^{\omega(d)} \sum_{c|d} 1 = (2A_{k,4})^{\omega(d)}$. This last sum is elementarily $\ll_k (\log y)^{A_{k,5}}$. Combining, we obtain

$$\sum_{1<d\leqslant y}^{\flat} \sum_{\boldsymbol{h}} \epsilon_{\boldsymbol{o} \cup \boldsymbol{h}}(d) \ll_{k,\mathscr{C}} y^{k-1} (\log y)^{A_{k,5}}. \tag{4.28}$$

Combining (4.24) with (4.25), (4.26), and (4.28) gives (4.18). $\qquad\square$

Remark 4.8 Montgomery and Soundararajan [21, Theorem 2 and (17)] give a very precise estimate for the left-hand side of (4.18) in the case where $\boldsymbol{o} = \varnothing$ and $\mathscr{C} = \{(x_1, \ldots, x_k) \in \mathbb{R}^k : 0 < x_1 < \cdots < x_k \leqslant 1\}$, viz., for any given $\epsilon > 0$,

$$\sum_{0<h_1<\cdots<h_k\leqslant y} \mathfrak{S}_{\boldsymbol{h}} = \frac{y^k}{k!} \left\{ 1 - \binom{k}{2}\frac{\log y}{y} - \binom{k}{2}\frac{\log(2\pi) + \gamma - 1}{y} + O_{k,\epsilon}\left(\frac{1}{y^{3/2-\epsilon}}\right) \right\}.$$

Thus, the error term in (4.18) can only be improved by a power of $\log y$. $\qquad\square$

5   Main Results

We establish each conclusion of Theorem 1.4 separately, on a weaker hypothesis than the one stated, viz. a certain *average* version of Conjecture 1.3. Let us introduce some notation in order to formulate our hypotheses. Given $x \geqslant 2$, a finite set $\boldsymbol{h} \subset \mathbb{Z}$, say $\boldsymbol{h} = \{h_1, \ldots, h_k\}$ where $h_1 < \cdots < h_k$, and any quantity X, let $\mathscr{R}_{\boldsymbol{h}}(x; X)$ be defined by the relation

$$\sum_{n\leqslant x} \mathbf{1}_{\mathbb{P}}(n + h_1) \cdots \mathbf{1}_{\mathbb{P}}(n + h_k) =: \mathfrak{S}_{\boldsymbol{h}} X + \mathscr{R}_{\boldsymbol{h}}(x; X), \tag{5.1}$$

where $\mathfrak{S}_{\boldsymbol{h}}$ is the singular series for $\boldsymbol{h}$ (defined in (1.1), and in (4.14), (4.15)). Given $\boldsymbol{k} = (k_1, \ldots, k_r) \in \mathbb{N}_0^r$ with $k_1 + \cdots + k_r = k$ and $\hat{\boldsymbol{\lambda}} = (\lambda_1, \ldots, \lambda_r) \in \mathbb{R}^r$, let

$$\mathscr{C}_{\hat{\boldsymbol{k}}, \hat{\boldsymbol{\lambda}}}^{+} := \{(x_1, \ldots, x_k) \in \Delta^k : (5.3) \text{ holds for } j = 1, \ldots, r\}, \tag{5.2}$$

where (5.3) is the condition

$$\lambda_1 + \cdots + \lambda_{j-1} < x_{k_1+\cdots+k_{j-1}+1} < \cdots < x_{k_1+\cdots+k_j} \leqslant \lambda_1 + \cdots + \lambda_j; \tag{5.3}$$

also, let

$$\mathscr{C}_{\hat{\boldsymbol{k}}, \hat{\boldsymbol{\lambda}}} := \{(x_1, \ldots, x_k) \in \Delta^k : x_{k_1+\cdots+k_j} - x_{k_1+\cdots+k_{j-1}} \leqslant \lambda_j, j = 1, \ldots, r\}. \tag{5.4}$$

In these definitions, and below, we set $\lambda_0 := 0$, $k_0 := 0$, and $x_0 := 0$. (Thus, for $j = 1, \lambda_1 + \cdots + \lambda_{j-1} = \lambda_0 = 0$, $k_1 + \cdots + k_{j-1} = k_0 = 0$, etc.)

Hypothesis $(o, r, \hat{\boldsymbol{\lambda}}, \hat{\boldsymbol{m}}, \hat{\boldsymbol{\ell}})$ Set $o := \varnothing$, or set $o := \{0\}$. Fix $r \in \mathbb{N}$, then fix $\hat{\boldsymbol{\lambda}} := (\lambda_1, \ldots, \lambda_r) \in \mathbb{R}_+^r$, $\hat{\boldsymbol{m}} := (m_1, \ldots, m_r) \in \mathbb{N}_0^r$, and $\hat{\boldsymbol{\ell}} := (\ell_1, \ldots, \ell_r) \in \mathbb{N}_0^r$. Let $\hat{\boldsymbol{k}} := (k_1, \ldots, k_r) := (\ell_1 + m_1, \ldots, \ell_r + m_r)$, let $k := k_1 + \cdots + k_r$, and let $\mathscr{C}_{\hat{k}, \hat{\lambda}}^+$ be as in (5.2), (5.3). Let x and y be parameters tending to infinity in such a way that $y \sim \log x$. As $x \to \infty$,

$$\sum_{(h_1, \ldots, h_k) \in y\mathscr{C}_{\hat{k}, \hat{\lambda}}^+ \cap \mathbb{Z}^k} \left| \mathscr{R}_{o \cup \boldsymbol{h}}(x; X_{\#o + k}) \right| = o\big(x/(\log x)^{\#o}\big),$$

where in the summand, $\boldsymbol{h} = \{h_1, \ldots, h_k\}$, $X_{\#o + k} \sim \int_2^x \mathrm{d}t/(\log t)^{\#o + k}$ as $x \to \infty$, and $\mathscr{R}_{o \cup \boldsymbol{h}}(x; X_{\#o + k})$ is as in (5.1).

Hypothesis $(\{0\}, r, \hat{\boldsymbol{\lambda}}, \hat{\boldsymbol{k}})$ Fix $r \in \mathbb{N}$, then fix $\hat{\boldsymbol{\lambda}} := (\lambda_1, \ldots, \lambda_r) \in \mathbb{R}_+^r$ and $\hat{\boldsymbol{k}} := (k_1, \ldots, k_r) \in \mathbb{N}^r$. Let $k := k_1 + \cdots + k_r$, and let $\mathscr{C}_{\hat{k}, \hat{\lambda}}$ be as in (5.4). Let x and y be parameters tending to infinity in such a way that $y \sim \log x$. As $x \to \infty$,

$$\sum_{(h_1, \ldots, h_k) \in y\mathscr{C}_{\hat{k}, \hat{\lambda}} \cap \mathbb{Z}^k} \left| \mathscr{R}_{\{0\} \cup \boldsymbol{h}}(x; X_{1 + k}) \right| = o(x/\log x),$$

where in the summand, $\boldsymbol{h} = \{h_1, \ldots, h_k\}$, $X_{1 + k} \sim \int_2^x \mathrm{d}t/(\log t)^{1 + k}$ as $x \to \infty$, and $\mathscr{R}_{\{0\} \cup \boldsymbol{h}}(x; X_{1 + k})$ is as in (5.1).

Theorem 5.1 Fix $r \in \mathbb{N}$, $\hat{\boldsymbol{\lambda}} := (\lambda_1, \ldots, \lambda_r) \in \mathbb{R}_+^r$, and $\hat{\boldsymbol{m}} := (m_1, \ldots, m_r) \in \mathbb{N}_0^r$. Let x and y be parameters tending to infinity in such a way that $y \sim \log x$.

(a) If Hypothesis $(\varnothing, r, \hat{\boldsymbol{\lambda}}, \hat{\boldsymbol{m}}, \hat{\boldsymbol{\ell}})$ holds for every fixed $\hat{\boldsymbol{\ell}} \in \mathbb{N}_0^r$, then, as $x \to \infty$,

$$\frac{1}{x} \sum_{\substack{n \leqslant x \\ \#\mathbb{P} \cap (n, n + (\lambda_1 + \cdots + \lambda_j)y] = m_1 + \cdots + m_j \\ j = 1, \ldots, r}} 1 \sim \prod_{j=1}^r \left(e^{-\lambda_j} \frac{\lambda_j^{m_j}}{m_j!} \right).$$

(b) If Hypothesis $(\{0\}, r, \hat{\boldsymbol{\lambda}}, \hat{\boldsymbol{m}}, \hat{\boldsymbol{\ell}})$ holds for every fixed $\hat{\boldsymbol{\ell}} \in \mathbb{N}_0^r$, then, as $x \to \infty$,

$$\frac{1}{\pi(x)} \sum_{\substack{p \leqslant x \\ \#\mathbb{P} \cap (p, p + (\lambda_1 + \cdots + \lambda_j)y] = m_1 + \cdots + m_j \\ j = 1, \ldots, r}} 1 \sim \prod_{j=1}^r \left(e^{-\lambda_j} \frac{\lambda_j^{m_j}}{m_j!} \right).$$

Theorem 5.2 Fix $r \in \mathbb{N}$ and $\hat{\boldsymbol{\lambda}} := (\lambda_1, \ldots, \lambda_r) \in \mathbb{R}_+^r$. Let x and y be parameters tending to infinity in such a way that $y \sim \log x$. If Hypothesis $(\{0\}, r, \hat{\boldsymbol{\lambda}}, \hat{\boldsymbol{k}})$ holds *for every fixed $\hat{\boldsymbol{k}} \in \mathbb{N}^r$, then, as $x \to \infty$,*

$$\frac{1}{\pi(x)} \sum_{\substack{p_n \leqslant x \\ p_{n+j} - p_{n+j-1} \leqslant \lambda_j y \\ j=1,\ldots,r}} 1 \sim \prod_{j=1}^{r} \left(1 - e^{-\lambda_j}\right).$$

Deduction of Theorem 5.1 This is simply a matter of combining Proposition 4.1 with Theorem 4.4, while keeping track of error terms, which we have conveniently hypothesized to be just what we need. We give the details for part (a), and leave part (b) to the reader.

Let $y_0 := 0$ and $y_j = (\lambda_1 + \cdots + \lambda_j)y$, $j = 1, \ldots, r$, so that $y_j - y_{j-1} = \lambda_j y$, $j = 1, \ldots, r$. Let $m \in \mathbb{N}_0$ be arbitrarily large, but fixed. Choose any $\ell_1, \ldots, \ell_r \in \mathbb{N}_0$ satisfying $\ell_1 + \cdots + \ell_r \leqslant 2m + 1$. We have

$$\sum_{\substack{y_{j-1} < h_1^{(j)} < \cdots < h_{\ell_j + m_j}^{(j)} \leqslant y_j \\ j=1,\ldots,r}} \sum_{n \leqslant x} \prod_{j=1}^{r} \mathbf{1}_{\mathbb{P}}\left(n + h_1^{(j)}\right) \cdots \mathbf{1}_{\mathbb{P}}\left(n + h_{\ell_j + m_j}^{(j)}\right)$$

$$= \sum_{\substack{y_{j-1} < h_1^{(j)} < \cdots < h_{\ell_j + m_j}^{(j)} \leqslant y_j \\ j=1,\ldots,r}} \left\{ \mathfrak{S}_{\boldsymbol{h}} X_{\sum_{j \leqslant r}(\ell_j + m_j)} + \mathcal{R}_{\boldsymbol{h}}\left(x; X_{\sum_{j \leqslant r}(\ell_j + m_j)}\right) \right\},$$

which, on Hypothesis $(\varnothing, r, \hat{\boldsymbol{\lambda}}, \hat{\boldsymbol{m}}, \hat{\boldsymbol{\ell}})$, is equal to

$$X_{\sum_{j \leqslant r}(\ell_j + m_j)} \left\{ \sum_{\substack{y_{j-1} < h_1^{(j)} < \cdots < h_{\ell_j + m_j}^{(j)} \leqslant y_j \\ j=1,\ldots,r}} \mathfrak{S}_{\boldsymbol{h}} \right\} + o(x),$$

which, by Theorem 4.4 and Lemma 4.6 (the sum is over $y\mathscr{C}_{\hat{\boldsymbol{k}}, \hat{\boldsymbol{\lambda}}}^{+} \cap \mathbb{Z}^k$, $\hat{\boldsymbol{k}} = (k_1, \ldots, k_r)$, $k_j = \ell_j + m_j$, $j = 1, \ldots, r$), is in turn equal to

$$X_{\sum_{j \leqslant r}(\ell_j + m_j)} \left\{ y^{\sum_{j \leqslant r}(\ell_j + m_j)} \prod_{j=1}^{r} \frac{\lambda_j^{\ell_j + m_j}}{(\ell_j + m_j)!} \right\} \left\{1 + o(1)\right\} + o(x).$$

We are assuming that $y \sim \log x$ and

$$X_{\sum_{j \leqslant r}(\ell_j + m_j)} \sim \int_2^x dt/(\log t)^{\sum_{j \leqslant r}(\ell_j + m_j)} \sim x(\log x)^{-\sum_{j \leqslant r}(\ell_j + m_j)}$$

$$\sim xy^{-\sum_{j \leqslant r}(\ell_j + m_j)}$$

as $x \to \infty$. Combining everything so far yields, on Hypothesis $(\varnothing, r, \hat{\boldsymbol{\lambda}}, \hat{\boldsymbol{m}}, \hat{\boldsymbol{\ell}})$,

$$\sum_{\substack{y_{j-1}<h_1^{(j)}<\cdots<h_{\ell_j+m_j}^{(j)}\leqslant y_j \\ j=1,\ldots,r}} \sum_{n\leqslant x}\prod_{j=1}^r \mathbf{1}_{\mathbb{P}}(n+h_1^{(j)})\cdots\mathbf{1}_{\mathbb{P}}(n+h_{\ell_j+m_j}^{(j)})=x\left\{\prod_{j=1}^r \frac{\lambda_j^{\ell_j+m_j}}{(\ell_j+m_j)!}\right\}\{1+o(1)\}.$$

Substituting this into the inequalities in Proposition 4.1 [with $\mathscr{B}=\mathbb{N}\cap[1,x]$ and a notational change $k_i \mapsto m_j$] and simplifying, we find, assuming Hypothesis $(\varnothing, r, \hat{\boldsymbol{\lambda}}, \hat{\boldsymbol{m}}, \hat{\boldsymbol{\ell}})$ holds whenever $\ell_1+\cdots+\ell_r \leqslant 2m+1$, that

$$\frac{1}{x}\sum_{\substack{n\leqslant x \\ \#\mathbb{P}\cap(n+y_{j-1},n+y_j]=m_j \\ j=1,\ldots,r}} 1 \geqslant \{1+o(1)\}\left\{\prod_{j=1}^r \frac{\lambda_j^{m_j}}{m_j!}\right\}\sum_{\substack{(\ell_1,\ldots,\ell_r)\in\mathbb{N}_0^r \\ \ell_1+\cdots+\ell_r\leqslant 2m+1}}\prod_{j=1}^r \frac{(-\lambda_j)^{\ell_j}}{\ell_j!}$$

and

$$\frac{1}{x}\sum_{\substack{n\leqslant x \\ \#\mathbb{P}\cap(n+y_{j-1},n+y_j]=m_j \\ j=1,\ldots,r}} 1 \leqslant \{1+o(1)\}\left\{\prod_{j=1}^r \frac{\lambda_j^{m_j}}{m_j!}\right\}\sum_{\substack{(\ell_1,\ldots,\ell_r)\in\mathbb{N}_0^r \\ \ell_1+\cdots+\ell_r\leqslant 2m}}\prod_{j=1}^r \frac{(-\lambda_j)^{\ell_j}}{\ell_j!},$$

from which we deduce that

$$\liminf_{x\to\infty}\frac{1}{x}\sum_{\substack{n\leqslant x \\ \#\mathbb{P}\cap(n+y_{j-1},n+y_j]=m_j \\ j=1,\ldots,r}} 1 \geqslant \left\{\prod_{j=1}^r \frac{\lambda_j^{m_j}}{m_j!}\right\}\sum_{\substack{(\ell_1,\ldots,\ell_r)\in\mathbb{N}_0^r \\ \ell_1+\cdots+\ell_r\leqslant 2m+1}}\prod_{j=1}^r \frac{(-\lambda_j)^{\ell_j}}{\ell_j!}$$

and

$$\limsup_{x\to\infty}\frac{1}{x}\sum_{\substack{n\leqslant x \\ \#\mathbb{P}\cap(n+y_{j-1},n+y_j]=m_j \\ j=1,\ldots,r}} 1 \leqslant \left\{\prod_{j=1}^r \frac{\lambda_j^{m_j}}{m_j!}\right\}\sum_{\substack{(\ell_1,\ldots,\ell_r)\in\mathbb{N}_0^r \\ \ell_1+\cdots+\ell_r\leqslant 2m}}\prod_{j=1}^r \frac{(-\lambda_j)^{\ell_j}}{\ell_j!}.$$

The sums over $\ell_1,\ldots,\ell_r$ are Taylor polynomial approximations to $e^{-(\lambda_1+\cdots+\lambda_r)}$. We have chosen m to be arbitrarily large, so we may conclude that, if Hypothesis $(\varnothing, r, \hat{\boldsymbol{\lambda}}, \hat{\boldsymbol{m}}, \hat{\boldsymbol{\ell}})$ holds for every fixed $\hat{\boldsymbol{\ell}} \in \mathbb{N}_0^r$,

$$\liminf_{x\to\infty}\frac{1}{x}\sum_{\substack{n\leqslant x\\ \#\mathbb{P}\cap(n+y_{j-1},n+y_j]=m_j\\ j=1,\ldots,r}}1\geqslant\prod_{j=1}^{r}\left(e^{-\lambda_j}\frac{\lambda_j^{m_j}}{m_j!}\right)\geqslant\limsup_{x\to\infty}\frac{1}{x}$$

$$\times\sum_{\substack{n\leqslant x\\ \#\mathbb{P}\cap(n+y_{j-1},n+y_j]=m_j\\ j=1,\ldots,r}}1,$$

and part (a) follows. As already mentioned, the proof of (b) is similar, but we take $\mathscr{B}=\mathbb{P}\cap[1,x]$ in Proposition 4.1. □

Deduction of Theorem 5.2 This is similar to the deduction of Theorem 5.1, but uses Proposition 4.2 (with $\mathscr{B}=\mathbb{P}\cap[1,x]$) instead of Proposition 4.1. We leave the details to the reader. □

6 Concluding Remarks

We believe that there exist more precise asymptotics for the distribution of primes in intervals than the ones stated in Conjectures 1.1 and 1.2. For instance, we believe that, for fixed $\lambda>0$ and $m\in\mathbb{N}_0$,

$$\sum_{\substack{n\leqslant x\\ \#\mathbb{P}\cap(n,n+\lambda\log x]=m}}1\sim x\left(e^{-\lambda}\frac{\lambda^m}{m!}\right)+\text{ lower order terms.}$$

We do not know exactly what shape these lower order terms should take. But perhaps, on a suitably strong version of the Hardy–Littlewood prime tuples conjecture, lower order terms may be obtained using more precise estimates, á la Montgomery and Soundararajan [21] (see Remark 4.8), for singular series averages. It may suffice to start with a k-tuples conjecture with uniformity k and an explicit error term, and to make all error terms in the argument explicit, with explicit dependence on k. We may pursue this in future work.

Acknowledgements The author is grateful to the anonymous referee, for carefully reading this manuscript and making corrections.

References

1. W.D. Banks, T. Freiberg, J. Maynard, On limit points of the sequence of normalized prime gaps. Proc. Lond. Math. Soc. 3, **113**, 515–539 (2016)
2. C. David, D. Koukoulopoulos, E. Smith, Sums of Euler products and statistics of elliptic curves. Math. Ann. **368**, 685–752 (2017)

3. K. Ford, Simple proof of Gallagher's singular series sum estimate (2011). Preprint. arXiv:1108.3861
4. T. Freiberg, Short intervals with a given number of primes. J. Number Theory **163**, 159–171 (2016)
5. P.X. Gallagher, On the distribution of primes in short intervals. Mathematika **23**, 4–9 (1976)
6. D.A. Goldston, J. Pintz, C.Y. Yıldırım, Primes in tuples I. Ann. Math. 2 **170**, 819–862 (2009)
7. A. Granville, P. Kurlberg, Poisson statistics via the Chinese remainder theorem. Adv. Math. **218**, 2013–2042 (2008)
8. G.H. Hardy, L.E. Littlewood, Some problems of 'Partitio numerorum'; III: on the expression of a number as a sum of primes. Acta Math. **44**, 1–70 (1923)
9. A. Hildebrand, H. Maier, Gaps between prime numbers. Proc. Am. Math. Soc. **104**, 1–9 (1988)
10. C. Hooley, On the difference between consecutive numbers prime to n. III Math. Z. **90**, 355–364 (1965)
11. C. Hooley, On the intervals between consecutive terms of sequences, in *Proceedings of the Symposium in Pure Mathematics of the American Mathematical Society, held at St. Louis University, St. Louis, MO, March 27–30, 1972*, ed. by H.G. Diamond. Proceedings of Symposia in Pure Mathematics, vol. XXIV (American Mathematical Society, Providence, RI, 1973), pp. 129–140
12. N.M. Katz, P. Sarnak, *Random Matrices, Frobenius Eigenvalues, and Monodromy*. American Mathematical Society Colloquium Publications, vol. 45 (American Mathematical Society, Providence, RI, 1999)
13. E. Kowalski, Averages of Euler products, distribution of singular series and the ubiquity of Poisson distribution. Acta Arith. **148**, 153–187 (2011)
14. P. Kurlberg, The distribution of spacings between quadratic residues. II. Isr. J. Math. **120**, 205–224 (2000)
15. P. Kurlberg, Poisson spacing statistics for value sets of polynomials. Int. J. Number Theory **5**, 489–513 (2009)
16. P. Kurlberg, Z. Rudnick, The distribution of spacings between quadratic residues. Duke Math. J. **100**, 211–242 (1999)
17. S. Lang, *Algebraic Number Theory*. Graduate Texts in Mathematics, 2nd edn., vol. 110 (Springer, New York, 1994)
18. H. Maier, Primes in short intervals. Mich. Math. J. **32**, 221–225 (1985)
19. J. Maynard, Small gaps between primes. Ann. Math. 2. **181**, 383–413 (2015)
20. J. Maynard, Dense clusters of primes in subsets. Compos. Math. **152**, 1517–1554 (2016)
21. H.L. Montgomery, K. Soundararajan, Primes in short intervals. Commun. Math. Phys. **252**, 589–617 (2004)
22. J. Pintz, On the singular series in the prime k-tuple conjecture (2010), Preprint. arXiv:1004.1084
23. J. Pintz, A note on the distribution of normalized prime gaps (2015). Preprint. arXiv:1510.04577
24. K. Soundararajan, The distribution of prime numbers, in *Equidistribution in Number Theory, An Introduction*, ed. by A. Granville, Z. Rudnick. NATO Science Series. Series II, Mathematics, Physics and Chemistry, vol. 237 (Springer, Dordrech, 2007), pp. 59–83
25. E. Westzynthius, Über die Verteilung der Zahlen, die zu den n ersten Primzahlen teilerfremd sind. Commentat. Phys.-Math. **5**, 1–37 (1931)

Distribution of Large Gaps Between Primes

Scott Funkhouser, Daniel A. Goldston, and Andrew H. Ledoan

Abstract We survey some past conditional results on the distribution of large gaps between consecutive primes and examine how the Hardy–Littlewood prime k-tuples conjecture can be applied to this question.

1 Introduction

The distribution of gaps between consecutive primes around their average spacing is expected to be a Poisson distribution. Thus, while at first glance the sequence of gaps appears random and irregular, we expect that they follow a very regular and well-behaved probability distribution. However, as we move to the distribution of larger than average gaps we expect to find increasing irregularity, especially as we reach the limiting size for these gaps. One would guess that near this maximal gap size any distribution will be exceedingly irregular. However, at present the available theoretical tools and well-accepted conjectures do not provide any widely believed standard model for large gaps.

Define

$$N(x, H) = \sum_{\substack{p_{n+1} \leq x \\ p_{n+1} - p_n \geq H}} 1 \qquad (1)$$

S. Funkhouser
Space and Naval Warfare Systems Center Atlantic, North Charleston, SC, USA

D. A. Goldston (✉)
Department of Mathematics and Statistics, San José State University, San José, CA, USA
e-mail: daniel.goldston@sjsu.edu

A. H. Ledoan
Department of Mathematics, University of Tennessee at Chattanooga, Chattanooga, TN, USA
e-mail: andrew-ledoan@utc.edu

© Springer International Publishing AG, part of Springer Nature 2018   45
J. Pintz, M. Th. Rassias (eds.), *Irregularities in the Distribution of Prime Numbers*,
https://doi.org/10.1007/978-3-319-92777-0_3

and the weighted counting function

$$S(x, H) = \sum_{\substack{p_{n+1} \leq x \\ p_{n+1} - p_n \geq H}} (p_{n+1} - p_n).$$ (2)

Gallagher [6] showed that a uniform version of the prime k-tuples conjecture of Hardy and Littlewood implies that the primes are distributed in a Poisson distribution around their average. In Sect. 2 we will discuss how this result implies that, for fixed $\lambda > 0$,

$$N(x, \lambda \log x) \sim e^{-\lambda} \frac{x}{\log x}, \qquad \text{as } x \to \infty,$$ (3)

and

$$S(x, \lambda \log x) \sim (1 + \lambda) e^{-\lambda} x, \qquad \text{as } x \to \infty.$$ (4)

These results are for fixed λ, but we are interested in larger gaps. One approach is to assume that the Hardy–Littlewood conjectures hold for primes up to x and for all k-tuples, where $k \leq f(x) \to \infty$ for some specified function $f(x)$, together with some strong error term. The Hardy–Littlewood conjectures will certainly fail when $k \asymp \log x$ and the error terms are often of size greater than $x^{1/2}$, and therefore this approach has definite limitations. There are also obstacles in applying these conjectures to $N(x, H)$ and $S(x, H)$. However, if we ignore these issues and consider this approach as only heuristic, then the following conjecture seems reasonable.

Poisson Tail Conjecture *For any $\epsilon > 0$ and $1 \leq H \leq \log^{2-\epsilon} x$, we have*

$$N(x, H) \asymp e^{-H/\log x} \frac{x}{\log x} \quad \text{and} \quad S(x, H) \asymp \left(1 + \frac{H}{\log x}\right) e^{-H/\log x} x.$$ (5)

For $H > \log^{2+\epsilon} x$, we have

$$N(x, H) = S(x, H) = 0.$$ (6)

Here, $f(x) \asymp g(x)$ means $f(x) \ll g(x)$ and $g(x) \ll f(x)$. In the critical range $\log^{2-\epsilon} x \leq H \leq \log^{2+\epsilon} x$, we have nothing to contribute. Other authors have made stronger conjectures than (6). In 1935 Cramér [3] conjectured that

$$\limsup_{p_n \to \infty} \frac{p_{n+1} - p_n}{\log^2 p_n} = 1,$$ (7)

while in 1995 Granville [11] conjectured that Cramér's conjecture is false and that

$$\limsup_{p_n \to \infty} \frac{p_{n+1} - p_n}{\log^2 p_n} \geq 2e^{-\gamma} = 1.12292\ldots, \tag{8}$$

based on a Cramér model modified to include divisibility by small primes. Our conjecture is much weaker and only implies that for any $\delta > 0$, there are prime gaps of size $> \log^{2-\delta} p_n$, and there are no prime gaps of size $> \log^{2+\delta} p_n$. It should be mentioned however that the same modification Granville used in the Cramér model was exploited by Maier [15] to prove there is no asymptotic formula for the number of primes in intervals $(x, x + \log^C x]$, for any given positive number C. This result of Maier demonstrates that all of these conjectures on large gaps are far from certain.

Our purpose in this paper is to describe earlier conditional work on large gaps between primes. The earliest such work assumes the Riemann Hypothesis and is mainly due to Cramér [4] and to Selberg [21]. Later, after Montgomery's work on the pair correlation of zeros of the Riemann zeta-function [16], some of these Riemann Hypothesis results were slightly improved by Gallagher and Mueller [7], by Mueller [19], and by Heath-Brown [14] assuming a pair correlation conjecture. All of these results can be obtained from estimating the second moment (or variance) of the number of primes in short intervals. As an application of these second moment results, one can obtain conditionally nearly optimal upper bounds on the sum

$$\mathscr{C}(x) = \sum_{p_{n+1} \leq x} (p_{n+1} - p_n)^2, \tag{9}$$

and also the closely related and slightly simpler sum

$$\mathscr{S}(x) = \sum_{p_{n+1} \leq x} \frac{(p_{n+1} - p_n)^2}{p_{n+1}}. \tag{10}$$

As far as we know, $\mathscr{C}(x)$ was first bounded on the Riemann Hypothesis by Cramér [4] in 1936, while $\mathscr{S}(x)$ was studied by Selberg [21] in 1943.

To obtain stronger results, we return to the Hardy–Littlewood conjectures. The conjecture for pairs (or two-tuples) with a strong error term is well-known to provide the same estimates for the second moment for primes in short intervals as that obtained by assuming the Riemann Hypothesis and the Pair Correlation Conjectures. In 2004 Montgomery and Soundararajan [18] were able to extend this method to give asymptotic formulas for the $2k$th moments for the primes in short intervals, assuming the Hardy–Littlewood conjecture for tuples of size $\leq 2k$ with a strong error term in the conjecture. At present, this approach is the most promising direction towards connecting results on large gap problems to a well-established, if extremely difficult, conjecture on primes. In Sect. 8 of this paper we use a fourth moment result to nearly resolve the conjectured asymptotic formulas for $\mathscr{C}(x)$ and $\mathscr{S}(x)$.

Notation We always assume that k, m, and n are integers. We denote the nth prime by p_n, and p will always denote a prime. By ϵ we mean any sufficiently small positive real number.

2 Gallagher's Theorem and the Poisson Distribution of Primes

We first introduce the Hardy–Littlewood prime k-tuples conjecture in the form used by Gallagher. Let $\mathcal{H}_k = \{h_1, \ldots, h_k\}$ be a set of k distinct nonnegative integers. Let $\pi(x; \mathcal{H}_k)$ denote the number of positive integers $n \le x$ for which $n + h_1, \ldots, n + h_k$ are simultaneously primes. Then the simplest form of the Hardy–Littlewood prime k-tuples conjecture [13] may be stated as follows.

Let

$$\mathfrak{S}(\mathcal{H}_k) = \prod_p \left(1 - \frac{1}{p}\right)^{-k} \left(1 - \frac{v_{\mathcal{H}_k}(p)}{p}\right), \tag{11}$$

where $v_{\mathcal{H}_k}(p)$ denotes the number of distinct residue classes modulo p occupied by the elements of $\mathcal{H}_k$. Note that, in particular, if $v_{\mathcal{H}_k}(p) = p$ for some prime p, then $\mathfrak{S}(\mathcal{H}_k) = 0$. However, if $v_{\mathcal{H}_k}(p) < p$ for all primes p, then $\mathfrak{S}(\mathcal{H}_k) \neq 0$ in which case the set $\mathcal{H}_k$ is called *admissible*.

Hardy–Littlewood Prime k-Tuples Conjecture *For each fixed integer $k \ge 2$ and admissible set $\mathcal{H}_k$, we have*

$$\pi(x; \mathcal{H}_k) = \mathfrak{S}(\mathcal{H}_k)\frac{x}{\log^k x}(1 + o_k(1)) \tag{12}$$

uniformly for $\mathcal{H}_k \subset [1, h]$, where $h \sim \lambda \log x$ as $x \to \infty$ and λ is a positive constant.

If $\mathcal{H}_k$ is not admissible, then there is a fixed prime p that always divides at least one of the k numbers $n + h_i$, with $1 \le i \le k$, and hence

$$\pi(x; \mathcal{H}_k) \le k \qquad \text{if } \mathcal{H}_k \text{ is not admissible.} \tag{13}$$

While a proof of (12) appears beyond our current state of knowledge, we do know by sieve methods (see [12]) the useful upper bound

$$\pi(x; \mathcal{H}_k) \ll_k \mathfrak{S}(\mathcal{H}_k)\frac{x}{\log^k x}. \tag{14}$$

Theorem 1 (Gallagher) *Let $P_k(N, h)$ denote the number of positive integers $n \le N$ for which the interval $(n, n + h]$ contains exactly k primes. Assuming the Hardy–*

Littlewood prime k-tuples conjecture, we have

$$P_k(N, h) \sim \frac{e^{-\lambda}\lambda^k}{k!}N, \quad for \ h \sim \lambda \log N \ as \ N \to \infty.$$

The Poisson distribution of primes manifests itself in Gallagher's proof through the fact that the singular series is on average asymptotic to 1 when averaged over all tuples. Gallagher [6] proved that, as $h \to \infty$,

$$\sum_{\substack{1 \leq h_1,\ldots,h_k \leq h \\ h_1,\ldots,h_k \ \text{distinct}}} \mathfrak{S}(\mathcal{H}_k) = h^k + O(h^{k-1/2+\epsilon}), \tag{15}$$

for each fixed $k \geq 2$.

Proof We give Gallagher's proof. For k a positive integer, the kth moment for the number of primes in the interval $(n, n+h]$ is

$$M_k(N) = \sum_{n \leq N}(\pi(n+h) - \pi(n))^k = \sum_{n \leq N}\sum_{n < p_1,\ldots,p_k \leq n+h} 1.$$

We group terms according to the number r of distinct primes among the primes $p_1, \ldots, p_k$ and obtain

$$M_k(N) = \sum_{r=1}^{k}\begin{Bmatrix} k \\ r \end{Bmatrix}\sum_{\substack{1 \leq h_1,\ldots,h_r \leq h \\ h_1,\ldots,h_r \ \text{distinct}}} \pi(N; \mathcal{H}_r),$$

where $\begin{Bmatrix} k \\ r \end{Bmatrix}$ is used to denote the Stirling number of the second type equal to the number of partitions of a set of k elements into r nonempty subsets. By (12), (13), and (15), we have for fixed k and $h \sim \lambda \log N$ as $N \to \infty$,

$$M_k(N) \sim \sum_{r=1}^{k}\begin{Bmatrix} k \\ r \end{Bmatrix}\sum_{\substack{1 \leq h_1,\ldots,h_r \leq h \\ h_1,\ldots,h_r \ \text{distinct}}} \mathfrak{S}(\mathcal{H}_r)\frac{N}{(\log N)^r}$$

$$\sim \sum_{r=1}^{k}\begin{Bmatrix} k \\ r \end{Bmatrix}h^r\frac{N}{(\log N)^r}$$

$$\sim m_k(\lambda)N,$$

where

$$m_k(\lambda) = \sum_{r=1}^{k}\begin{Bmatrix} k \\ r \end{Bmatrix}\lambda^r, \tag{16}$$

which is the kth Poisson moment with expected value λ. Theorem 1 now follows from the standard theorems on moments.

From Theorem 1, we now prove (3) and (4).

Theorem 2 *Assuming the Hardy–Littlewood prime k-tuples conjecture, then for fixed $\lambda > 0$ and $H \sim \lambda \log x$ as $x \to \infty$, we have*

$$N(x, \lambda \log x) \sim e^{-\lambda} \frac{x}{\log x} \tag{17}$$

and

$$S(x, \lambda \log x) \sim (1 + \lambda)e^{-\lambda}x. \tag{18}$$

Proof Let

$$S_1(x, H) = \sum_{\substack{p_n \leq x \\ p_{n+1} - p_n \geq H}} ((p_{n+1} - p_n) - H). \tag{19}$$

Taking $k = 0$ in Theorem 1, we have

$$P_0(N, h) \sim e^{-\lambda}N,$$

where $P_0(N, h)$ is the number of $j \leq N$ for which the interval $(j, j+h]$ contains no primes. This interval has the same number of primes as the interval $[j+1, j+\lfloor h \rfloor]$, which contains no primes if and only if there is an n for which $p_n \leq j$ and $p_{n+1} \geq j + \lfloor h \rfloor + 1$, which can occur if and only if $p_{n+1} - p_n \geq \lfloor h \rfloor + 1$. Hence, in this case, $p_n \leq j \leq p_{n+1} - \lfloor h \rfloor - 1$, and there are $p_{n+1} - p_n - \lfloor h \rfloor$ such j's for this p_n. Thus, we have

$$P_0(N, h) = \sum_{\substack{p_n \leq N \\ p_{n+1} - p_n \geq \lfloor h \rfloor}} (p_{n+1} - p_n - \lfloor h \rfloor)$$

$$= \sum_{\substack{p_n \leq N \\ p_{n+1} - p_n \geq h}} (p_{n+1} - p_n - \lfloor h \rfloor)$$

$$= S_1(N, h) + O\left(\frac{N}{\log N}\right).$$

We conclude that, for $H \sim \lambda \log x$,

$$S_1(x, H) \sim e^{-\lambda} x. \tag{20}$$

From (1), (2), and (19),

$$S_1(x, H) = S(x, H) - HN(x, H),$$

so that (18) follows from (17) and (20). To prove (17), we note that

$$S_1(x, H) = \int_H^\infty N(x, u)\, du \tag{21}$$

and, since $N(x, u)$ is a nonincreasing function of u, we have for any $\delta > 0$,

$$\frac{1}{\delta H} \int_H^{(1+\delta)H} N(x, u)\, du \leq N(x, H) \leq \frac{1}{\delta H} \int_{(1-\delta)H}^H N(x, u)\, du.$$

Therefore, $N(x, H)$ is bounded between

$$\frac{S_1(x, H) - S_1(x, (1 \pm \delta)H)}{\pm \delta H},$$

which by (20) is, as $\delta \to 0$,

$$\sim \left(\frac{e^{-\lambda} - e^{-(1 \pm \delta)\lambda}}{\pm \delta \lambda} \right) \frac{x}{\log x}$$

$$\sim \left(\frac{1 - e^{\mp \delta \lambda}}{\pm \delta \lambda} \right) e^{-\lambda} \frac{x}{\log x}$$

$$\sim (1 + O(\delta \lambda)) e^{-\lambda} \frac{x}{\log x}$$

$$\sim e^{-\lambda} \frac{x}{\log x},$$

thus proving (17).

There is an alternative approach for proving Theorem 2 which avoids moments. In [9] the second two authors proved, using inclusion-exclusion with the Hardy–Littlewood prime k-tuples conjecture, that for fixed $\lambda > 0$ and $d \sim \lambda \log x$,

$$\mathcal{N}(x, d) = \sum_{\substack{p_{n+1} \leq x \\ p_{n+1} - p_n = d}} 1 \sim e^{-\lambda} \mathfrak{S}(d) \frac{x}{\log^2 x}, \tag{22}$$

where

$$
\mathfrak{S}(d) = \begin{cases} 2C_2 \displaystyle\prod_{\substack{p \mid d \\ p > 2}} \left(\dfrac{p-1}{p-2} \right) & \text{if } d \text{ is even,} \\[2em] 0 & \text{if } d \text{ is odd,} \end{cases}
$$

and

$$
C_2 = \prod_{p>2} \left(1 - \frac{1}{(p-1)^2} \right) = 0.66016\ldots .
$$

Here, $\mathfrak{S}(d)$ is the singular series given in (11) when $k = 2$ and $\mathcal{H}_2 = \{0, d\}$. As a consequence of (15) we have

$$
\sum_{d \le h} \mathfrak{S}(d) \sim h.
$$

Hence, with $H \sim \lambda \log x$, by partial summation,

$$
\sum_{\substack{p_{n+1} \le x \\ p_{n+1} - p_n < H}} 1 = \sum_{d < H} \mathcal{N}(x, d)
$$

$$
\sim \left(\sum_{d < H} e^{-d/\log x} \mathfrak{S}(d) \right) \frac{x}{\log^2 x}
$$

$$
\sim \left(\int_0^{\lambda \log x} e^{-u/\log x} \, du \right) \frac{x}{\log^2 x}
$$

$$
\sim \left(1 - e^{-\lambda} \right) \frac{x}{\log x}.
$$

Thus,

$$
N(x, H) = \sum_{p_n \le x} 1 - \sum_{\substack{p_{n+1} \le x \\ p_{n+1} - p_n < H}} 1
$$

$$
\sim e^{-\lambda} \frac{x}{\log x},
$$

which proves (3). The same argument gives

$$S(x, H) = \sum_{p_n \leq x} (p_{n+1} - p_n) - \sum_{\substack{p_{n+1} \leq x \\ p_{n+1} - p_n < H}} (p_{n+1} - p_n)$$

$$\sim x - \sum_{d < H} d \mathcal{N}(x, d)$$

$$\sim x - \left(\int_0^{\lambda \log x} u e^{-u/\log x} \, du \right) \frac{x}{\log^2 x}$$

$$\sim (1 + \lambda) e^{-\lambda} x,$$

which proves (18).

3   Bounding the Number of Large Gaps with Moments

Moving to larger gaps between primes, we reduce our goal of finding their distribution and only seek to find bounds on their frequency. A simple method for bounding the number of large gaps was introduced by Selberg [21]. Let

$$M_{2k}(x, h) = \int_1^x (\vartheta(y + h) - \vartheta(y) - h)^{2k} \, dy, \tag{23}$$

where

$$\vartheta(x) = \sum_{p \leq x} \log p$$

and k is a positive integer.

Lemma 1 *For $k \geq 1$ and $H \geq 1$, we have*

$$S(x, H) \ll \left(\frac{2}{H} \right)^{2k} M_{2k} \left(x, \frac{H}{2} \right). \tag{24}$$

Proof We have $\vartheta(y + h) - \vartheta(y) = 0$ whenever there are consecutive primes $p_n \leq y$ and $y + h < p_{n+1}$, which is when $y \in [p_n, p_{n+1} - h)$ which has length $(p_{n+1} - p_n) - h$. Suppose $p_{n+1} - p_n \geq H$ and take $h = H/2$. Then

$$\int_{p_n}^{p_{n+1} - h} (\vartheta(y + h) - \vartheta(y) - h)^{2k} \, dy = h^{2k}((p_{n+1} - p_n) - h) \geq \frac{1}{2} h^{2k}(p_{n+1} - p_n).$$

Summing over $p_{n+1} \leq x$ gives the result.

There is a slightly different moment often used in this subject, namely

$$m_{2k}(x, h) = \int_1^x (\psi(y + h) - \psi(y) - h)^{2k} \, dy, \tag{25}$$

where

$$\psi(x) = \sum_{p^m \le x} \log p = \sum_{n \le x} \Lambda(n).$$

The next lemma shows that the two moments are essentially the same size. The error term here saves only a power of $\log x$ over the actual size of these moments, but that is sufficient for our applications.

Lemma 2 *For $k \ge 1$, $x \ge 2$, and $h \ge 1$, we have*

$$m_{2k}(x, h)^{1/2k} = M_{2k}(x, h)^{1/2k} + O\left(\left(xh^k\right)^{1/2k}\right).$$

Proof We recall Minkowski's inequality for integrals (see [17])

$$\left(\int_a^b |f(x) + g(x)|^p \, dx\right)^{1/p} \le \left(\int_a^b |f(x)|^p \, dx\right)^{1/p} + \left(\int_a^b |g(x)|^p \, dx\right)^{1/p}.$$

Since

$$\psi(y + h) - \psi(y) - h = (\vartheta(y + h) - \vartheta(y) - h) + R(y, h),$$

where

$$R(y, h) = \sum_{\substack{y < p^m \le y+h \\ m \ge 2}} \log p,$$

we obtain from Minkowski's inequality

$$m_{2k}(x, h)^{1/2k} = M_{2k}(x, h)^{1/2k} + O\left(\left(\int_1^x R(y, h)^{2k} \, dy\right)^{1/2k}\right). \tag{26}$$

It remains to estimate $R(y, h)$. The inequality $y < p^m \le y + h$ is equivalent to $\sqrt[m]{y} < p \le \sqrt[m]{y+h}$. We make use of the inequality $\sqrt[m]{y+h} \le \sqrt[m]{y} + \sqrt[m]{h}$ when h is large and the inequality $\sqrt[m]{y+h} \le \sqrt[m]{y}(1 + h/my)$ when h is small.

We consider first the case when $h \ge x^\delta$, for some fixed $\delta > 0$. Using the sieve bound

$$\pi(y + H) - \pi(y) \ll \frac{H}{\log H},$$

we have for $1 \le y \le x$

$$R(y, h) \le \sum_{\substack{\sqrt[m]{y} < p \le \sqrt[m]{y} + \sqrt[m]{h} \\ m \ge 2}} \log p$$

$$\ll \sum_{2 \le m \le \log x} \frac{m \sqrt[m]{h}}{\log h} \log x$$

$$\ll \frac{\sqrt{h} \log x}{\log h} + \frac{\sqrt[3]{h} \log^3 x}{\log h}$$

$$\ll \sqrt{h}.$$

Substituting this bound into (26) proves Lemma 2 in this range.

Next, we consider the range $1 \le h \le x^\delta$. Estimating trivially, we have

$$R(y, h) \le \sum_{\substack{\sqrt[m]{y} < p \le \sqrt[m]{y} + h / (my^{1 - 1/m}) \\ m \ge 2}} \log p \ll \frac{h \log^2 x}{y^{1/2}}.$$

Hence,

$$\int_1^x R(y, h)^{2k} \, dy \ll h^{2k} (\log x)^{4k+1} \ll x^{2k\delta} (\log x)^{4k+1} \ll x,$$

on taking $\delta = 1/4k$. This proves Lemma 2 in this range.

4 Second Moment Results Assuming the Riemann Hypothesis

In [21] Selberg proved[1] that, assuming the Riemann Hypothesis, for $T \ge 2$,

$$\int_1^{T^4} \left(\vartheta \left(y + \frac{y}{T} \right) - \vartheta(y) - \frac{y}{T} \right)^2 \frac{dy}{y^2} \ll \frac{\log^2 T}{T}. \tag{27}$$

The left-hand side, here, is a damped second moment for primes in short intervals where the interval length varies as a fixed multiple of where it is located. We will make use of (27) in Sect. 5. Most authors use in place of Selberg's second moment either $M_2(x, h)$ or $m_2(x, h)$. Saffari and Vaughan [20] found a method for going back and forth between moments using fixed intervals $[x, x+h]$ and moments using

---

[1] Selberg also proved an unconditional estimate that we are not concerned with in this paper.

intervals $(x, x + \delta x]$. (See, also, [10].) The result corresponding to (27) is, for $1 \leq h \ll x^{3/4}$,

$$M_2(x, h) \ll hx \log^2 x. \tag{28}$$

This result may also be proved directly using the explicit formula. (See [5] and [20].)

Theorem 3 (Selberg) *Assuming the Riemann Hypothesis, we have for $H > 0$*

$$S(x, H) = \sum_{\substack{p_{n+1} \leq x \\ p_{n+1} - p_n \geq H}} (p_{n+1} - p_n) \ll \frac{x}{H} \log^2 x. \tag{29}$$

Proof Taking $k = 1$ in Lemma 1, we obtain Theorem 3 with the additional condition that $H \ll x^{3/4}$. From (29),

$$N(x, H) \leq \frac{1}{H} S(x, H) \ll \frac{x}{H^2} \log^2 x.$$

Now, if $H \geq Cx^{1/2} \log x$, we can take C sufficiently large to obtain $N(x, H) < 1$. Therefore, for a sufficiently large constant,

$$N(x, H) = S(x, H) = 0, \qquad \text{if } H \geq Cx^{1/2} \log x. \tag{30}$$

Thus, we may drop the condition $H \ll x^{3/4}$ in Theorem 3, since the better estimate (30) holds in this range.[2]

The result (30) implies the following result of Cramér [2] from 1920.

Corollary 1 (Cramér) *Assuming the Riemann Hypothesis, we have*

$$p_{n+1} - p_n \ll \sqrt{p_n} \log p_n.$$

We also have

Corollary 2 (Selberg) *Assuming the Riemann Hypothesis, we have*

$$\mathscr{C}(x) = \sum_{p_{n+1} \leq x} (p_{n+1} - p_n)^2 \ll x \log^3 x.$$

Proof Since

$$\mathscr{C}(x) = \int_0^x S(x, H) \, dH \leq x + \int_1^x S(x, H) \, dH, \tag{31}$$

the result follows from (29).

---

[2]Recent work [1] has determined that $C = 0.84$ is acceptable.

5   Selberg's Result on $\mathscr{S}(x)$

There is a further result from Selberg's original paper that deserves special mention.

Theorem 4 (Selberg) *Assuming the Riemann Hypothesis, we have*

$$\mathscr{S}(x) = \sum_{p_{n+1} \leq x} \frac{(p_{n+1} - p_n)^2}{p_{n+1}} \ll \log^3 x.$$

This is only a single power of $\log x$ larger than the conjectured size of $\mathscr{S}(x)$, while the result for $\mathscr{C}(x)$ obtained in the previous section is two powers of $\log x$ larger than the conjecture. (See (43) in Sect. 8.) We will see in the next section that, assuming a pair correlation conjecture for zeros of the Riemann zeta-function we can recover a $\log x$ in the results of Sect. 8. However, this is not true for Theorem 4, where assuming a pair correlation conjecture does not give any improvement.

From this theorem, we easily obtain the following corollary which is partly in the direction of Corollary 4 proved in the next section assuming a pair correlation conjecture.

Corollary 3 *Assuming the Riemann Hypothesis, we have*

$$\liminf_{x \to \infty} \frac{\mathscr{C}(x)}{x \log^2 x} \ll 1. \tag{32}$$

Proof From the identity

$$\mathscr{S}(x) = \frac{\mathscr{C}(x)}{x} + \int_1^x \frac{\mathscr{C}(u)}{u^2} \, du, \tag{33}$$

we see that, if $\mathscr{C}(u) \geq Cu \log^2 u$ for $\sqrt{x} \leq u \leq x$,

$$\mathscr{S}(x) > \int_{\sqrt{x}}^x \frac{\mathscr{C}(u)}{u^2} \, du \geq C \int_{\sqrt{x}}^x \frac{\log^2 u}{u} \, du = \frac{7}{24} C \log^3 x,$$

which contradicts Theorem 4 if C is large enough.

Proof (of Theorem 4) We return to (27) and follow Selberg's proof. We take x large, let $1 \leq H \leq x^{3/4}$, and take $T = 2x/H$. Then (27) becomes

$$\int_1^x \left(\vartheta \left(y + \frac{Hy}{2x} \right) - \vartheta(y) - \frac{Hy}{2x} \right)^2 \frac{dy}{y^2} \ll \frac{H}{x} \log^2 x. \tag{34}$$

Suppose now that $p_{n+1} \leq x$ and $p_{n+1} - p_n \geq (H/x)p_{n+1}$. Then, just as in Lemma 1, we have

$$\int_{p_n}^{p_{n+1}-(H/x)p_{n+1}} \left(\vartheta \left(y + \frac{Hy}{2x} \right) - \vartheta(y) - \frac{Hy}{2x} \right)^2 \frac{dy}{y^2} = \int_{p_n}^{p_{n+1}-(H/x)p_{n+1}} \frac{H^2}{4x^2} \, dy$$

$$\geq \frac{H^2}{8x^2}(p_{n+1} - p_n).$$

Hence, for $H \leq x^{3/4}$ we obtain the slight refinement of Theorem 3 that, assuming the Riemann Hypothesis,

$$\sum_{\substack{p_{n+1} \leq x \\ p_{n+1}-p_n \geq (H/x)p_{n+1}}} (p_{n+1} - p_n) \ll \frac{x}{H} \log^2 x.$$

The condition $H \leq x^{3/4}$ may be dropped in view of Cramér's bound (40), and Theorem 4 now follows on integrating with respect to H from 1 to x.

6 Second Moment Results Assuming the Riemann Hypothesis and Pair Correlation

The results in the last section assuming the Riemann Hypothesis have never been improved. However, in 1972 Montgomery [16] found an additional conjecture on the vertical distribution of zeros of the Riemann zeta-function which allows us to obtain essentially the best possible second moment results. The Riemann Hypothesis states that the complex zeros of the Riemann zeta-function have their real part equal to $1/2$, so that a complex zero can be written as $\rho = 1/2 + i\gamma$, where γ is real. If this conjecture is false, then the primes will have a dramatically more irregular behavior than we expect. However, all evidence points to the truth of the Riemann Hypothesis, but without pointing towards a method for its proof. Montgomery introduced the function, for $T \geq 2$,

$$F(\alpha) = \frac{1}{N(T)} \sum_{0 < \gamma, \gamma' \leq T} T^{i\alpha(\gamma-\gamma')} \omega(\gamma - \gamma'), \tag{35}$$

where $\omega(u) = 4/(4 + u^2)$, and

$$N(T) = \sum_{0 < \gamma \leq T} 1 = \frac{T}{2\pi} \log \frac{T}{2\pi e} + O(\log T). \tag{36}$$

Montgomery's Theorem *Assume the Riemann Hypothesis. For any real α we have $F(\alpha)$ is even, $F(\alpha) \geq 0$, and for $0 \leq \alpha \leq 1$ we have*

$$F(\alpha) = T^{-2\alpha} \log T(1 + o(1)) + \alpha + o(1), \qquad as \ T \to \infty. \tag{37}$$

This theorem determines $F(\alpha)$ for $|\alpha| \leq 1$, while for larger α Montgomery made the following conjecture.

Montgomery's Conjecture *We have*

$$F(\alpha) = 1 + o(1) \qquad for\ 1 < \alpha \leq M, \qquad as\ T \to \infty, \tag{38}$$

for any fixed number M.

The connection between this conjecture and the second moment for primes is given in the following theorem [10] from 1987.

Equivalence Theorem *Assuming the Riemann Hypothesis, then Montgomery's conjecture is equivalent to*

$$m_2(x, h) \sim hx \log \frac{x}{h} \tag{39}$$

uniformly for $1 \leq h \leq x^{1-\epsilon}$.

In all except one of the applications to large gaps between primes we only need a weaker conjecture than (38), which we shall state as follows.

Bounded $F(\alpha)$ Conjecture *For any $\delta > 0$, we have $F(\alpha) \ll 1$ uniformly for $1 \leq \alpha \leq 2 + \delta$.*

Heath-Brown [14] proved, assuming the Riemann Hypothesis and the Bounded $F(\alpha)$ Conjecture, that for $1 \leq h \leq x^{1/2+\delta}$

$$m_2(x, h) \ll hx \log x.$$

Using Lemma 2, the same result stated above also holds when we replace $m_2(x, h)$ by $M_2(x, h)$.

Therefore, we obtain the following results obtained in the same way as the corresponding results proved in the last section.

Theorem 5 (Heath-Brown) *Let $x \geq 2$ and $H \geq 1$. Assuming the Riemann Hypothesis and the Bounded $F(\alpha)$ Conjecture, we have*

$$N(x, H) \ll \frac{x}{H^2} \log x \quad and \quad S(x, H) \ll \frac{x}{H} \log x.$$

Corollary 4 *Assuming the Riemann Hypothesis and the Bounded $F(\alpha)$ Conjecture, we have*

$$p_{n+1} - p_n \ll \sqrt{p_n \log p_n}. \tag{40}$$

We also have

Corollary 5 *Assuming the Riemann Hypothesis and the Bounded $F(\alpha)$ Conjecture, we have*

$$\mathscr{C}(x) = \sum_{p_{n+1} \leq x} (p_{n+1} - p_n)^2 \ll x \log^2 x.$$

Montgomery's Conjecture and the Riemann Hypothesis also give the slightly stronger result that

$$p_{n+1} - p_n = o(\sqrt{p_n \log p_n}). \tag{41}$$

This was first proved in [8]. It is also an easy consequence of (39), which implies, for $y = y(x) = o(x)$, that

$$M_2(x + y, h) - M_2(x, h) = o(hx \log x).$$

7 Montgomery and Soundararajan's Higher Moment Results Using a Strong Prime k-Tuples Conjecture

In place of $\pi(x; \mathscr{H}_k)$, we now make use of

$$\psi(x; \mathscr{H}_k) = \sum_{n \leq x} \Lambda(n + h_1) \cdots \Lambda(n + h_k), \tag{42}$$

which has the advantage of counting primes with a constant density of one. A strong form of the Hardy–Littlewood Conjecture now takes the following form.

Strong Hardy–Littlewood Prime k-Tuples Conjecture *For a fixed integer $k \geq 2$ and admissible set $\mathscr{H}_k$, we have for any $\epsilon > 0$ and x sufficiently large*

$$\psi(x; \mathscr{H}_k) = \mathfrak{S}(\mathscr{H}_k)x + O(x^{1/2+\epsilon}),$$

uniformly for $\mathscr{H}_k \subset [1, h]$.

The following theorem is a special case of the main theorem in [18].

Theorem 6 (Montgomery–Soundararajan) *Suppose the Strong Hardy–Littlewood Prime k-Tuples Conjecture holds for $2 \leq k \leq 2K$ and uniformly for $\mathscr{H}_k \subset [1, h]$. Then for $x \geq 2$ and $\log x \leq h \leq x^{1/2K}$, we have*

$$m_{2K}(x, h) = h^K x \log^K \frac{x}{h} \left(1 + o(1) + O\left(\left(\frac{h}{\log x} \right)^{-1/16K} \right) \right) + O(h^{2K} x^{1/2+\epsilon}).$$

If $K = 1$ we recover the asymptotic formula for $m_2(x, h)$ in (39) for a restricted range of h. By Lemma 2, we see Theorem 6 also holds for $M_{2K}(x, h)$. Hence, we obtain the bound

$$M_{2K}(x, h) \ll h^K x \log^K \frac{x}{h}$$

for $\log x \leq h \leq x^{1/2K-\delta}$ for any fixed $K \geq 1$ and $\delta > 0$. Hence, by Lemma 1 we obtain the following result on large gaps.

Theorem 7 *Suppose that the Strong Hardy–Littlewood Prime k-Tuples Conjecture holds for $2 \leq k \leq 2K$ and uniformly for $\mathcal{H}_k \subset [1, H]$. Then for $x \geq 2$ and $\log x \leq H \leq x^{1/2K-\delta}$ for any fixed $\delta > 0$, we have*

$$S(x, H) \ll \frac{x}{H^K} \log^K x.$$

This result is consistent with the Poisson Tail Conjecture (5) but, of course, much weaker.

8 Application of the Fourth Moment Bound to $\mathcal{C}(x)$ and $\mathcal{S}(x)$

We expect that

$$\mathcal{C}(x) \sim 2x \log x \quad \text{and} \quad \mathcal{S}(x) \sim \log^2 x \quad \text{as } x \to \infty. \tag{43}$$

We see from (33) that the asymptotic formula for $\mathcal{C}(x)$ immediately implies the asymptotic formula for $\mathcal{S}(x)$, and therefore we concentrate on $\mathcal{C}(x)$.

By combining Gallagher's Theorem with the fourth moment bound in Theorem 7, we are able to nearly evaluate $\mathcal{C}(x)$. The result we obtain is the following theorem.

Theorem 8 *Assume the Hardy–Littlewood Prime k-Tuples Conjecture as in (12) and the Strong Hardy–Littlewood k-Tuples Conjecture for $2 \leq k \leq 4$ uniformly for $\mathcal{H}_k \subset [1, x^{1/4-\delta}]$ for any fixed $\delta > 0$. We have*

$$\mathcal{C}(x) = 2x \log x (1 + o(1)) + O\left(\sum_{\substack{p_{n+1} \leq x \\ p_{n+1} - p_n \geq x^{1/4-\delta}}} (p_{n+1} - p_n)^2 \right).$$

We do not expect any prime gaps as large as those in the error term here. However, the existence of a single prime gap of size $\sqrt{p_n \log p_n}$ in $[x/2, x]$ would invalidate the asymptotic formula for $\mathcal{C}(x)$. The Strong Hardy–Littlewood

Conjecture without some additional information on how the error terms average when combined cannot disprove the existence of such long gaps. As we have seen in (41), the pair correlation conjecture can (barely) show such gaps do not exist, but that conjecture is equivalent to a second moment results on primes in an extended range.

Proof (of Theorem 8) As in (31), we have

$$\mathscr{C}(x) = \int_0^x S(x, H) \, dH$$

$$= \left(\int_0^{\lambda_0 \log x} + \int_{\lambda_0 \log x}^{\lambda_1 \log x} + \int_{\lambda_1 \log x}^{x^{1/4-\delta}} + \int_{x^{1/4-\delta}}^x \right) S(x, H) \, dH$$

$$= I_1 + I_2 + I_3 + I_4.$$

Here, we let $\lambda_0 \to 0$ and $\lambda_1 \to \infty$ sufficiently slowly. Since $S(x, H) \le x$,

$$I_1 \le \lambda_0 x \log x = o(x \log x).$$

By Theorem 2,

$$I_2 = (1 + o(1))x \log x \int_{\lambda_0}^{\lambda_1} (1 + \lambda)e^{-\lambda} \, d\lambda$$

$$= (1 + o(1))x \log x \left(e^{-\lambda_0}(2 + \lambda_0) - e^{-\lambda_1}(2 + \lambda_1) \right)$$

$$= (1 + o(1))2x \log x.$$

Applying Theorem 7 with $K = 2$, we obtain

$$I_3 \ll \int_{\lambda_1 \log x}^{x^{1/4-\delta}} \frac{x \log^2 x}{H^2} \, dH \ll \frac{x \log x}{\lambda_1} = o(x \log x).$$

Finally,

$$I_4 = \sum_{\substack{p_{n+1} \le x \\ p_{n+1} - p_n \ge x^{1/4-\delta}}} (p_{n+1} - p_n) \int_{x^{1/4-\delta}}^{p_{n+1}-p_n} dH \le \sum_{\substack{p_{n+1} \le x \\ p_{n+1} - p_n \ge x^{1/4-\delta}}} (p_{n+1} - p_n)^2.$$

This completes the proof of Theorem 8.

9 Some Numerical Results on Large Gaps

In this section we present some numerical studies related to the behaviors addressed in this paper. It is instructive first to recall how the largest gap between primes no greater than x increases with x. Figure 1 is a plot of the maximal gap

$$g^*(x) = \sup_{p_{n+1} \leq x} (p_{n+1} - p_n),$$

along with the analytical asymptotic form

$$\widetilde{g}^*(x) = \log^2 x$$

advanced by Cramér [4] over a representative sampling of approximately logarithmically spaced prime x.

Next, we consider the large-gap counting function $N(x, H)$. For convenience, let us define the expected asymptotic form in (3) as $\widetilde{N}(x, \lambda \log x) = e^{-\lambda}x/\log x$. Figures 2, 3, and 4 are logarithmic plots of $N(x, \lambda \log x)$ along with $\widetilde{N}(x, \lambda \log x)$ for $\lambda = 1, 3,$ and 6, respectively.

The weighted analogue of $N(x, H)$ is $S(x, H)$ from (2), and its expected asymptotic form in (4) is defined here as $\widetilde{S}(x, \lambda \log x) = (1 + \lambda)e^{-\lambda}x$ for $H = \lambda \log x$. Figures 5, 6, and 7 are logarithmic plots of $S(x, \lambda \log x)$ along with $\widetilde{S}(x, \lambda \log x)$ for $\lambda = 1, 3,$ and 6.

The relative errors in $\widetilde{N}(x, \lambda \log x)$ and $\widetilde{S}(x, \lambda \log x)$ decrease with increasing x, for a given λ, in support of the conjectured asymptotic behaviors. We also find, however, that the relative error increases with increasing λ for a fixed x. We may interpret this behavior as being consistent with the expected non-Poissonian properties in the distribution of large gaps.

Fig. 1 $g^*(x)$ and $\widetilde{g}^*(x)$ plotted in asterisks and circles, respectively, for a representative sampling of prime x

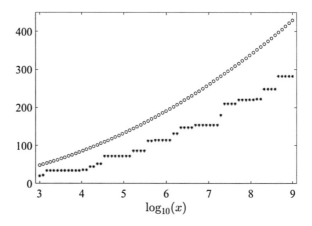

Fig. 2 $\log_{10} N(x, \log x)$ and $\log_{10} \widetilde{N}(x, \log x)$ plotted in asterisks and circles, respectively, for a representative sampling of prime x

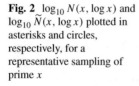

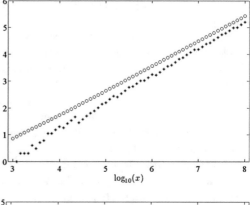

Fig. 3 $\log_{10} N(x, 3\log x)$ and $\log_{10} \widetilde{N}(x, 3\log x)$ plotted in asterisks and circles, respectively, for a representative sampling of prime x

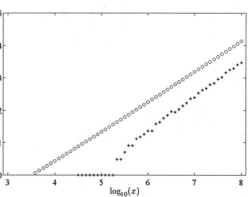

Fig. 4 $\log_{10} N(x, 6\log x)$ and $\log_{10} \widetilde{N}(x, 6\log x)$ plotted in asterisks and circles, respectively, for a representative sampling of prime x

Finally, let us consider the terms $\mathscr{C}(x)$ and $\mathscr{S}(x)$ and the expected behaviors articulated in (43). For convenience, we define $\widetilde{\mathscr{C}}(x) = 2x\log x$ to represent the expected asymptotic form of $\mathscr{C}(x)$. Curiously, the asymptotic form of $\mathscr{S}(x)$ is identical to the Cramér's maximal gap bound $\widetilde{g}^*(x)$. Figure 8 is a logarithmic plot of $\mathscr{C}(x)$ along with $\widetilde{\mathscr{C}}(x)$. Figure 9 is a logarithmic plot of $\mathscr{S}(x)$ along with $\widetilde{g}^*(x)$.

Fig. 5 $\log_{10} S(x, \log x)$ and $\log_{10} \widetilde{S}(x, \log x)$ plotted in asterisks and circles, respectively, for a representative sampling of prime x

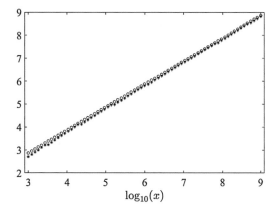

Fig. 6 $\log_{10} S(x, 3 \log x)$ and $\log_{10} \widetilde{S}(x, 3 \log x)$ plotted in asterisks and circles, respectively, for a representative sampling of prime x

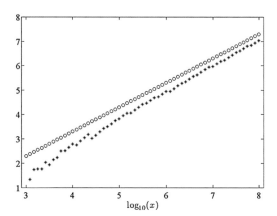

Fig. 7 $\log_{10} S(x, 6 \log x)$ and $\log_{10} \widetilde{S}(x, 6 \log x)$ plotted in asterisks and circles, respectively, for a representative sampling of prime x

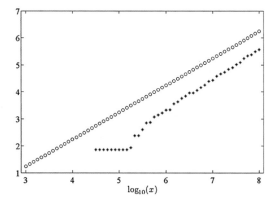

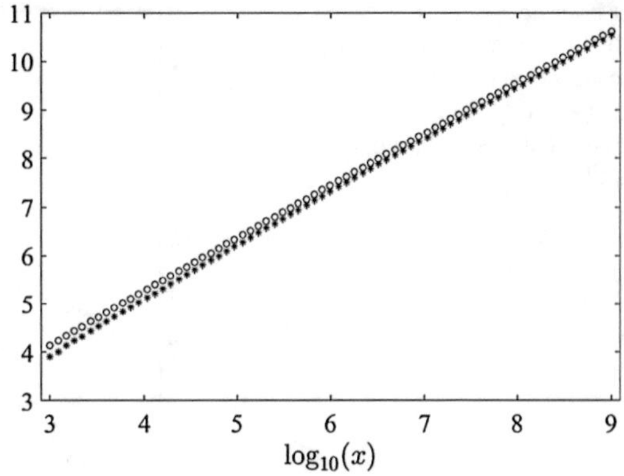

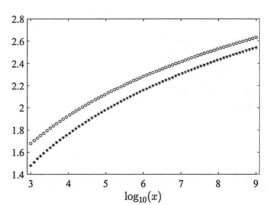

Fig. 8 $\log_{10} \mathscr{C}(x)$ and $\log_{10} \widetilde{\mathscr{C}}(x)$ plotted in asterisks and circles, respectively, for a representative sampling of prime x

Fig. 9 $\log_{10} \mathscr{S}(x)$ and $\log_{10} \widetilde{g}^*(x)$ plotted in asterisks and circles, respectively, for a representative sampling of prime x

Acknowledgements The authors wish to express their sincere gratitude and appreciation to the anonymous referee for carefully reading the original version of this paper and for making a number of very helpful comments and suggestions.

References

1. E. Carneiro, M.B. Milinovich, K. Soundararajan, Fourier optimization and prime gaps (submitted for publication). Available at https://arxiv.org/abs/1708.04122
2. H. Cramér, Some theorems concerning prime numbers. Arkiv för Mat. Astr. och Fys. **15**(5), 1–32 (1920)
3. H. Cramér, Prime numbers and probability. Skand. Mat. Kongr. **8**, 107–115 (1935)
4. H. Cramér, On the order of magnitude of the difference between consecutive prime numbers. Acta Arith. **2**, 23–46 (1936)
5. P.X. Gallagher, Some consequences of the Riemann hypothesis. Acta Arith. **37**, 339–343 (1980)
6. P.X. Gallagher, On the distribution of primes in short intervals. Mathematika **23**, 4–9 (1976); Corrigendum: Mathematika **28**(1), 86 (1981)

7. P.X. Gallagher, J.H. Mueller, Primes and zeros in short intervals. J. Reine Angew. Math. **303/304**, 205–220 (1978)
8. D.A. Goldston, D.R. Heath-Brown, A note on the difference between consecutive primes. Math. Annalen **266**, 317–320 (1984)
9. D.A. Goldston, A.H. Ledoan, *On the Differences Between Consecutive Prime Numbers, I, Integers 12B* (2012/13), 8 pp. [Paper No. A3]. Also appears in: *Combinatorial Number Theory (Proceedings of the "Integers Conference 2011," Carrollton, Georgia, October 26–29, 2011), De Gruyter Proceedings in Mathematics* (2013), pp. 37–44
10. D.A. Goldston, H.L. Montgomery, Pair correlation of zeros and primes in short intervals, in *Analytic Number Theory and Diophantine Problems* (Birkhauser, Boston, MA, 1987), pp. 183–203
11. A. Granville, Harald Cramér and the distribution of prime numbers, Harald Cramér Symposium (Stockholm, 1993). Scand. Actuar. J. **1**, 12–28 (1995)
12. H. Halberstam, H.-E. Richert, *Sieve Methods.* London Mathematical Society Monographs, vol. 4 (Academic Press, London/New York/San Francisco, 1974)
13. G.H. Hardy, J.E. Littlewood, Some problems of 'Partitio numerorum'; III: on the expression of a number as a sum of primes. Acta Math. **44**(1), 1–70 (1922). Reprinted in *Collected Papers of G.H. Hardy*, vol. I (including joint papers with J.E. Littlewood and others; edited by a committee appointed by the London Mathematical Society) (Clarendon Press/Oxford University Press, Oxford, 1966), pp. 561–630
14. D.R. Heath-Brown, Gaps between primes, and the pair correlation of zeros of the zeta-function. Acta Arith. **41**, 85–99 (1982)
15. H. Maier, Primes in short intervals. Mich. Math. J. **32**(2), 221–225 (1985)
16. H.L. Montgomery, The pair correlation of zeros of the zeta function, in *Analytic Number Theory, Proceedings of Symposia in Pure Mathematics*, vol. XXIV (St. Louis University, St. Louis, MO, 1972), pp. 181–193; (American Mathematical Society, Providence, RI, 1973)
17. H.L. Montgomery, *Early Fourier Analysis.* Pure and Applied Undergraduate Texts, vol. 22 (American Mathematical Society, Providence, RI, 2014)
18. H.L. Montgomery, K. Soundararajan, Primes in short intervals. Commun. Math. Phys. **252**(1–3), 589–617 (2004)
19. J.H. Mueller, On the difference between consecutive primes, in *Recent Progress in Analytic Number Theory*, Durham, vol. 1 (1979), pp. 269–273 (Academic, London/New York, 1981)
20. B. Saffari, R.C. Vaughan, On the fractional parts of x/n and related sequences, II. Ann. Inst. Fourier (Grenoble) **27**(2), 1–30 (1977)
21. A. Selberg, On the normal density of primes in small intervals, and the difference between consecutive primes. Arch. Math. Naturvid. **47**(6), 87–105 (1943)

On the Difference in Values of the Euler Totient Function Near Prime Arguments

Stephan Ramon Garcia and Florian Luca

Abstract We prove unconditionally that for each $\ell \geqslant 1$, the difference $\varphi(p - \ell) - \varphi(p + \ell)$ is positive for 50% of odd primes p and negative for 50%.

1 Introduction

In what follows, p always denotes an odd prime number. The inequality $\varphi(p-1) \geqslant \varphi(p+1)$ appears to hold for an overwhelming majority of twin primes p, $p+2$, and to be reversed for small, but positive, proportion of the twin primes [4]. To be more specific, if the Bateman–Horn conjecture is true, then the inequality above holds for at least 65.13% of twin prime pairs and is reversed for at least 0.47% of pairs. Numerical evidence suggests, in fact, that the ratio is something like 98% to 2%. In other words, for an overwhelming majority of twin prime pairs p, $p + 2$, it appears that the first prime has more primitive roots than does the second.

Based upon numerical evidence, it was conjectured in [4] that this bias disappears if only p is assumed to be prime. That is, $\varphi(p - 1) > \varphi(p + 1)$ for 50% of primes and the inequality is reversed for 50% of primes. We prove this unconditionally and, moreover, we are able to handle wider spacings as well. If all primality assumptions are dropped, then it is known that $\varphi(n - 1) > \varphi(n + 1)$ asymptotically 50% of the time. This follows from work of Shapiro, who considered the distribution function of $\varphi(n)/\varphi(n - 1)$ [6].

S. R. Garcia
Department of Mathematics, Pomona College, Claremont, CA, USA
e-mail: stephan.garcia@pomona.edu
http://pages.pomona.edu/~sg064747

F. Luca (✉)
School of Mathematics, University of the Witwatersrand, Johannesburg, South Africa
Max Planck Institute for Mathematics, Bonn, Germany

Department of Mathematics, Faculty of Sciences, University of Ostrava, Ostrava, Czech Republic
e-mail: Florian.Luca@wits.ac.za

© Springer International Publishing AG, part of Springer Nature 2018
J. Pintz, M. Th. Rassias (eds.), *Irregularities in the Distribution of Prime Numbers*,
https://doi.org/10.1007/978-3-319-92777-0_4

Table 1 The number (#) of primes $p \leqslant 2{,}038{,}074{,}743$ (the hundred millionth prime) for which $\varphi(p - \ell) = \varphi(p + \ell)$.

ℓ	#	ℓ	#	ℓ	#	ℓ	#	ℓ	#	ℓ	#	ℓ	#	ℓ	#
1	103	9	359	17	106	25	6	33	338	41	109	49	38	57	295
2	49	10	4	18	219	26	47	34	47	42	322	50	5	58	54
3	**201,078**	11	107	19	104	27	357	35	3	43	121	51	371	59	127
4	58	12	214	20	3	28	17	36	374	44	39	52	38	60	538
5	5	13	98	21	403	29	117	37	97	45	486	53	126	61	126
6	231	14	7	22	52	30	507	38	45	46	47	54	303	62	45
7	43	**15**	**108,772**	23	136	31	98	39	380	47	124	55	2	**63**	**22,654**
8	50	16	39	24	301	32	53	40	5	48	236	56	6	64	48

This number is exceptionally large if $\ell = 4^n - 1$; see Theorem 2 for an explanation

Let $\pi(x)$ denote the number of primes at most x and let $\sim$ denote asymptotic equivalence. The Prime Number Theorem ensures that $\pi(x) \sim x/\log x$. Our main theorem is the following.

Theorem 1 *Let ℓ be a positive integer. As $x \to \infty$ we have the following:*

(a) $\#\{p \leqslant x : \varphi(p - \ell) > \varphi(p + \ell)\} \sim \frac{1}{2}\pi(x).$
(b) $\#\{p \leqslant x : \varphi(p - \ell) < \varphi(p + \ell)\} \sim \frac{1}{2}\pi(x).$
(c) $\#\{p \leqslant x : \varphi(p - \ell) = \varphi(p + \ell)\} = o(\pi(x)).$

A curious phenomenon occurs in (c), in the sense that the decay rate relative to $\pi(x)$ depends upon ℓ in a peculiar manner. Theorem 2 shows that

$$\#\{p \leqslant x : \varphi(p - \ell) = \varphi(p + \ell)\} \ll \begin{cases} \dfrac{x}{(\log x)^3} & \text{if } \ell = 4^n - 1, \\[2mm] \dfrac{x}{e^{(\log x)^{1/3}}} & \text{otherwise.} \end{cases}$$

This does not appear to be an artifact of the proof since it is borne out in numerical computations (see Table 1) and is consistent with the Bateman–Horn conjecture.

We first prove Theorem 1 in the case $\ell = 1$. This is undertaken in Sect. 2 and it comprises the bulk of this article. For the sake of readability, we break the proof into several steps which we hope are easy to follow. In Sect. 3, we outline the modifications necessary to treat the case $\ell \geqslant 2$. This approach permits us to focus on the main ingredients that are common to both cases, without getting sidetracked by all of the adjustments necessary to handle the general case.

2 Proof of Theorem 1 for $\ell = 1$

2.1 The Case of Equality

Our first job is to show that the set of primes p for which $\varphi(p-1) = \varphi(p+1)$ has a counting function that is $o(\pi(x))$. We need the following lemma, which generalizes earlier work by Erdős, Pomerance, and Sárközy [2] in the case $k = 1$. The upper bound (b) in the following was strengthened in a preprint of Yamada [13].

Lemma 1 (Graham–Holt–Pomerance [5]) *Suppose that j and $j + k$ have the same prime factors. Let $g = \gcd(j, j + k)$ and suppose that*

$$\frac{jt}{g} + 1 \quad and \quad \frac{(j+k)t}{g} + 1 \tag{1}$$

are primes that do not divide j.

(a) Then $n = j\left(\dfrac{(j+k)t}{g} + 1\right)$ satisfies $\varphi(n) = \varphi(n+k)$.
(b) For k fixed and sufficiently large x, the number of solutions $n \leqslant x$ to $\varphi(n) = \varphi(n+k)$ that are not of the form above is less than $x/\exp((\log x)^{1/3})$.

Consider the case $k = 2$ and $n = p - 1$, in which p is prime. Suppose that j and $j + 2$ have the same prime factors and let $g = \gcd(j, j + 2)$. Let us also suppose that t is a positive integer such that

$$\frac{jt}{g} + 1 \quad and \quad \frac{(j+2)t}{g} + 1$$

are primes and

$$p - 1 = j\left(\frac{(j+2)t}{g} + 1\right).$$

Since j and $j + 2$ have the same prime factors, they are both powers of 2. Then $j = 2$ and $j + 2 = 4$, so $g = 2$. Consequently,

$$t + 1, \qquad 2t + 1, \quad and \quad 4t + 3 \tag{2}$$

are prime. Reduction modulo 3 reveals that at least one of them is a multiple of 3. The only prime triples produced by (2) are $(2, 3, 7)$ and $(3, 5, 11)$, in which $r = 1$ and $r = 2$, respectively. Consequently,

$$\#\{p \leqslant x : \varphi(p-1) = \varphi(p+1)\} < \frac{x}{\exp((\log x)^{1/3})} + 2 = o(\pi(x)). \tag{3}$$

This is Theorem 1.c in the case $\ell = 1$.

2.2 A Comparison Lemma

Instead of comparing $\varphi(p-1)$ and $\varphi(p+1)$ directly, it is more convenient to compare the related quantities

$$\frac{\varphi(p-1)}{p-1} = \prod_{q\mid(p-1)} \left(1 - \frac{1}{q}\right) \quad \text{and} \quad \frac{\varphi(p+1)}{p+1} = \prod_{q\mid(p+1)} \left(1 - \frac{1}{q}\right), \tag{4}$$

in which q is prime. Let

$$S(p) := \frac{\varphi(p-1)}{p-1} - \frac{\varphi(p+1)}{p+1}, \tag{5}$$

which we claim is nonzero for $p \geqslant 5$. Let $P(n)$ denote the largest prime factor of n. Since

$$\frac{\varphi(n)}{n} = \prod_{q\mid n} \left(\frac{q-1}{q}\right), \tag{6}$$

it follows that $P(n)$ is the largest prime factor of the denominator of $\varphi(n)/n$. Since $\gcd(p-1, p+1) = 2$, the condition $S(p) = 0$ implies that $p-1$ and $p+1$ are both powers of 2. Thus, $S(p) = 0$ holds only for $p = 3$.

Something similar to the following lemma is in [4], although there it was assumed that $p + 2$ is also prime. The adjustment for $\ell \geqslant 2$ is discussed in Sect. 3.

Lemma 2 (Comparison Lemma) *The set of primes p for which $\varphi(p-1) - \varphi(p+1)$ and $S(p)$ have the same sign has counting function asymptotic to $\pi(x)$ as $x \to \infty$.*

Proof In light of (3), it suffices to show that

$$\varphi(p-1) > \varphi(p+1) \quad \Longleftrightarrow \quad \frac{\varphi(p-1)}{p-1} > \frac{\varphi(p+1)}{p+1} \tag{7}$$

on a set of full density in the primes. The forward direction is clear, so we focus on the reverse. If the inequality on the right-hand side of (7) holds, then

$$0 < p\big(\varphi(p-1) - \varphi(p+1)\big) + \varphi(p-1) + \varphi(p+1)$$
$$\leqslant p\big(\varphi(p-1) - \varphi(p+1)\big) + \tfrac{1}{2}(p-1) + \tfrac{1}{2}(p+1)$$
$$= p\big(\varphi(p-1) - \varphi(p+1) + 1\big) \tag{8}$$

because $p-1$ and $p+1$ are even. Since $\varphi(n)$ is even for $n \geqslant 3$, it follows that $\varphi(p-1) - \varphi(p+1) \geqslant 0$. Now appeal to (3) to see that strict inequality holds on a set of full density in the primes.

2.3 Some Preliminaries

In our later study of the quantity $S(p)$, we need to avoid four classes of inconvenient primes. To make the required estimates, we need some notation. Let x be large, let $y := \log \log x$, and define

$$L_y := \mathrm{lcm}\{m : m \leqslant y\}. \tag{9}$$

Then $L_y = e^{\psi(y)}$, in which

$$\psi(y) := \sum_{p^k \leqslant y} \log p$$

is Chebyshev's function. Since the Prime Number Theorem asserts that $\psi(y) = y + o(y)$ as $y \to \infty$, we obtain

$$L_y = e^{y+o(y)} < e^{2y} = (\log x)^2 \tag{10}$$

for sufficiently large x. For a positive integer n, let $D_y(n)$ denote the largest divisor of n that is y-smooth:

$$D_y(n) := \max \{d : d|n \text{ and } P(d) \leqslant y\}. \tag{11}$$

On occasion, we will need the Brun sieve. Let $f_1, f_2, \ldots, f_m$ be a collection of distinct irreducible polynomials with positive leading coefficients. An integer n is *prime generating* for this collection if each $f_1(n), f_2(n), \ldots, f_m(n)$ is prime. Let $G(x)$ denote the number of prime-generating integers at most x and suppose that $f = f_1 f_2 \cdots f_m$ does not vanish identically modulo any prime. As $x \to \infty$,

$$G(x) \ll \frac{x}{(\log x)^m},$$

in which the implied constant depends only upon m and $\prod_{i=1}^{m} \deg f_i$ [12, Thm. 3, Sect. I.4.2]. The upper bound obtained in this manner has the same order of magnitude as the prediction furnished by the Bateman–Horn conjecture [1].

2.4 Inconvenient Primes of Type 1

Let

$$\mathscr{E}_1(x) := \{p \leqslant x : D_y(p^2 - 1) \nmid L_y\}.$$

We will prove that

$$\#\mathscr{E}_1(x) \ll \frac{x}{y^{1/2} \log x} = o(\pi(x)). \tag{12}$$

Suppose that $p \in \mathscr{E}_1(x)$. Then (11) and the definition of $\mathscr{E}_1(x)$ yield a divisor d of $p^2 - 1$ such that $P(d) \leqslant y$ and $d \nmid L_y$. These conditions provide a prime power q^b with least exponent b such that

$$q^b | d, \qquad q = P(q^b) \leqslant y, \quad \text{and} \quad y < q^b. \tag{13}$$

Indeed, if every y-smooth prime power q^b that divides d satisfies $q^b \leqslant y$, then (9) would imply that $d | L_y$, a contradiction. We also observe that the second two conditions in (13) ensure that $b \geqslant 2$.

We claim that for large x either $p - 1$ or $p + 1$ has a prime power divisor q^c with $c \geqslant 2$ in the interval $[y/2, y^2]$. Since

$$p^2 - 1 = (p - 1)(p + 1) \quad \text{and} \quad \gcd(p - 1, p + 1) = 2,$$

it follows that $d/(2, d)$ divides one of $p+1$, $p-1$. There are two cases to consider.

- If $q = 2$, then 2^{b-1} divides one of $p - 1$, $p + 1$. Since we aim to prove (12) as $x \to \infty$, we may assume that $b \geqslant 3$ since for $b = 2$, the third statement in (13) implies $\log \log x < 4$. Next observe that (13) implies that $y/2 < 2^{b-1}$. The minimality of b in (13) ensures that $2^{b-1} \leqslant y < y^2$. Thus, $2^{b-1} \in [\frac{y}{2}, y^2]$ has $b - 1 \geqslant 2$ (since $b \geqslant 3$), and divides one of $p - 1$, $p + 1$.
- If q is odd, then q^b divides $p - 1$ or $p + 1$. The minimality of b ensures that $q^{b-1} \leqslant y$ and the second statement in (13) yields

$$\frac{y}{2} < y < q^b = q^{b-1}q \leqslant y^2, \quad \text{and hence} \quad q^b \in [\tfrac{y}{2}, y^2] \quad \text{with } b \geqslant 2.$$

For large x, we conclude that one of $p - 1$, $p + 1$ has a prime power divisor q^c with $c \geqslant 2$ in $[y/2, y^2]$.

Let $\pi_s(x)$ denote the number of prime powers p^a with $a \geqslant 2$ that are at most x. Since $p^a \leqslant x$ with $a \geqslant 2$ implies either $a = 2$ and $p \leqslant x^{1/2}$, or $a \in [3, (\log x / \log 2)]$ and $p \leqslant x^{1/3}$, the Prime Number Theorem implies that

$$\pi_s(x) = \pi(\sqrt{x}) + O\big(\pi(x^{1/3}) \log x\big) = (2 + o(1)) \frac{x^{1/2}}{\log x}$$

as $x \to \infty$. Let $\pi(x; m, k)$ denote the number of primes at most x that are congruent to k modulo m. Then the Brun sieve implies

$$\#\mathcal{E}_1(x) \leqslant \sum_{\substack{q^b \in [y/2, y^2] \\ b \geqslant 2}} \pi(x; q^b, 1) + \sum_{\substack{q^b \in [y/2, y^2] \\ b \geqslant 2}} \pi(x; q^b, -1)$$

$$\ll \sum_{\substack{q^b \in [y/2, y^2] \\ b \geqslant 2}} \frac{x}{\varphi(q^b) \log x} \ll \sum_{\substack{q^b \in [y/2, y^2] \\ b \geqslant 2}} \frac{2x}{q^b \log x}$$

$$\ll \frac{x}{\log x} \sum_{\substack{q^b \in [y/2, y^2] \\ b \geqslant 2}} \frac{1}{q^b}$$

$$\ll \frac{x}{\log x} \int_{y/2}^{y^2} \frac{d\pi_s(t)}{t} \leqslant \frac{x}{\log x} \int_{y/2}^{\infty} \frac{d\pi_s(t)}{t}$$

$$\ll \frac{x}{\log x} \left(\left. \frac{\pi_s(t)}{t} \right|_{y/2}^{\infty} + \int_{y/2}^{\infty} \frac{\pi_s(t)\, dt}{t^2} \right)$$

$$\ll \frac{x}{\log x} \left(\frac{(y/2)^{1/2}}{(y/2) \log(y/2)} + \int_{y/2}^{\infty} t^{-3/2} (\log t)^{-1/2}\, dt \right)$$

$$\ll \frac{x}{y^{1/2} (\log y)(\log x)} = o(\pi(x))$$

as $x \to \infty$. This is the desired estimate (12).

2.5   Inconvenient Primes of Type 2

Fix a large x and define the function

$$h_y(n) := \sum_{\substack{r \mid n \\ r > y}} \frac{1}{r},$$

in which r is prime. Let

$$\mathcal{E}_2(x) := \left\{ p \leqslant x : h_y(p-1) > \frac{1}{y\sqrt{\log y}} \quad \text{or} \quad h_y(p+1) > \frac{1}{y\sqrt{\log y}} \right\}.$$

We claim that

$$\#\mathcal{E}_2(x) \ll \frac{\pi(x)}{\sqrt{\log y}} = o(\pi(x)) \tag{14}$$

as $x \to \infty$. This will follow from an averaging argument similar to [7, Lem. 3].
   The Brun sieve with $f(t) = rt \pm 1$ provides

$$\pi(x; r, \pm 1) \ll \frac{\pi(x)}{r} \qquad \text{for } y \leqslant r \leqslant (\log x)^3$$

uniformly for r in the specified range [12, Thm. 3, Sect. I.4.2]. We use the trivial estimate

$$\pi(x; r, \pm 1) \leqslant \frac{x}{r} \qquad \text{for } (\log x)^3 \leqslant r \leqslant x.$$

We also require the upper bound

$$\sum_{r \geqslant y} \frac{1}{r^2} = \int_y^\infty \frac{d\pi(t)}{t^2} = \frac{\pi(t)}{t^2} \bigg|_y^\infty + 2 \int_y^\infty \frac{\pi(t)\,dt}{t^3} \ll \frac{\pi(y)}{y^2} \ll \frac{1}{y \log y}, \qquad (15)$$

which is afforded by the Prime Number Theorem. As $x \to \infty$, we have

$$\sum_{p \leqslant x} h_y(p \pm 1) = \sum_{p \leqslant x} \sum_{\substack{r \mid (p \pm 1) \\ r > y}} \frac{1}{r} = \sum_{y \leqslant r \leqslant x} \frac{1}{r} \sum_{\substack{p \leqslant x \\ r \mid (p \pm 1)}} 1$$

$$= \sum_{y \leqslant r \leqslant x} \frac{\pi(x; r, \mp 1)}{r}$$

$$\ll \sum_{y \leqslant r \leqslant (\log x)^3} \frac{\pi(x)}{r^2} + \sum_{(\log x)^3 < r \leqslant x} \frac{x}{r^2}$$

$$\ll \pi(x) \sum_{r \geqslant y} \frac{1}{r^2} + x \sum_{r \geqslant (\log x)^3} \frac{1}{r^2}$$

$$\ll \frac{\pi(x)}{y \log y} + \frac{x}{(\log x)^3} \ll \frac{\pi(x)}{y \log y} + \frac{\pi(x)}{(\log x)^2}$$

$$\ll \frac{\pi(x)}{y \log y}.$$

Consequently,

$$\frac{\#\mathcal{E}_2(x)}{y\sqrt{\log y}} \leqslant \sum_{\substack{p \leqslant x \\ h_y(p-1) > \frac{1}{y\sqrt{\log y}}}} h_y(p-1) + \sum_{\substack{p \leqslant x \\ h_y(p+1) > \frac{1}{y\sqrt{\log y}}}} h_y(p+1)$$

$$\leqslant \sum_{p \leqslant x} h_y(p-1) + \sum_{p \leqslant x} h_y(p+1)$$

$$\ll \frac{\pi(x)}{y \log y},$$

which implies (14).

2.6 *Inconvenient Primes of Type 3*

Let $\omega(n)$ denote the number of distinct prime factors of n and $\omega_y(n)$ the number of distinct prime factors $q \leqslant y$ of n. Define

$$\mathscr{E}_3(x) = \left\{ p \leqslant x : \omega_y(p^2 - 1) \notin [1.5 \log \log y, 2.5 \log \log y] \right\}.$$

We claim that

$$\#\mathscr{E}_3(x) \ll \frac{\pi(x)}{\log \log y} = o(\pi(x)) \tag{16}$$

as $x \to \infty$. Although this is essentially a result of Erdős [3], we sketch a simpler proof that is easily generalized since we later need to handle $p^2 - \ell^2$ instead of $p^2 - 1$.

If $p \in \mathscr{E}_3(x)$, then for large x we have

$$\omega_y(p^2 - 1) + 1 = \omega_y(p - 1) + \omega_y(p + 1)$$

because $\gcd(p - 1, p + 1) = 2$. Thus, either

$$\min\{\omega_y(p - 1), \omega_y(p + 1)\} \leqslant 0.75 \log \log y + 1 \leqslant 0.8 \log \log y$$

or

$$\max\{\omega_y(p - 1), \omega_y(p + 1) > 1.25 \log \log y > 1.2 \log \log y$$

for sufficiently large x. Without loss of generality, we may suppose that

$$\omega_y(p - 1) \leqslant 0.8 \log \log y \qquad \text{or} \qquad \omega_y(p - 1) \geqslant 1.2 \log \log y. \tag{17}$$

Then

$$0.04(\log \log y)^2 \leqslant \left(\omega_y(p - 1) - \log \log y \right)^2 \tag{18}$$

and similarly if $p + 1$ occurs in (17).

We next require the following "Turán–Kubilius"-type result; see [8, Lem. 2], [10, §V.5, 1, p. 159]. To study $\varphi(p \pm \ell)$ for $\ell \neq 1$ requires a slight generalization; see Lemma 5 in Sect. 3 for a statement and sketch of the proof.

Lemma 3 (Motohashi) $\displaystyle\sum_{p \leqslant x} (\omega_y(p \pm 1) - \log \log y)^2 = O(\pi(x) \log \log y).$

Now return to (18), apply Lemma 3, and conclude that

$$0.04(\log\log y)^2 \#\mathscr{E}_3(x) \leqslant \sum_{p\in\mathscr{E}_3(x)} (\omega_y(p-1) - \log\log y)^2$$

$$+ (\omega_y(p+1) - \log\log y)^2$$

$$\leqslant \sum_{p\leqslant x} (\omega_y(p-1) - \log\log y)^2 + (\omega_y(p+1) - \log\log y)^2$$

$$= O(\pi(x)\log\log y).$$

This yields the desired estimate (16).

2.7 Inconvenient Primes of Type 4

Let

$$\mathscr{E}_4(x) = \left\{ p \leqslant x : \frac{p^2 - 1}{\varphi(p^2 - 1)} > (\log y)^{1/3} \right\}. \tag{19}$$

We claim that

$$\#\mathscr{E}_4(x) \ll \frac{\pi(x)}{(\log y)^{1/6}} = o(\pi(x)) \tag{20}$$

as $x \to \infty$. Since $\varphi(p-1)\varphi(p+1) \leqslant \varphi(p^2-1)$, the condition

$$\frac{p^2 - 1}{\varphi(p^2 - 1)} > (\log y)^{1/3}$$

implies that

$$\frac{p+1}{\varphi(p+1)} > (\log y)^{1/6} \quad \text{or} \quad \frac{p-1}{\varphi(p-1)} > (\log y)^{1/6}.$$

A standard application of the Siegel–Walfisz theorem yields

$$\sum_{p\leqslant x} \frac{p-1}{\varphi(p-1)} \ll \pi(x); \tag{21}$$

see [10, §I.28, 1b, p 30] or [9]. The same holds with $p - 1$ replaced by $p + 1$; the adjustments necessary to handle $p \pm \ell$ are discussed in Sect. 3. Thus,

$$\mathscr{E}_4(x)(\log y)^{1/6} \leqslant \sum_{p \in \mathscr{E}_4(x)} \left(\frac{p-1}{\varphi(p-1)} + \frac{p+1}{\varphi(p+1)} \right)$$

$$\leqslant \sum_{p \leqslant x} \left(\frac{p-1}{\varphi(p-1)} + \frac{p+1}{\varphi(p+1)} \right)$$

$$\ll \pi(x),$$

which yields (20).

2.8   Convenient Primes

Throughout the remainder of the proof, we let $5 \leqslant p \leqslant x$, in which x is large, and we suppose that

$$p \notin \mathscr{E}_1(x) \cup \mathscr{E}_2(x) \cup \mathscr{E}_3(x) \cup \mathscr{E}_4(x).$$

We say that such a prime is *convenient*. Because $\gcd(p-1, p+1) = 2$, we have

$$D_y(p^2 - 1) = m_1 m_2, \tag{22}$$

in which

$$p - 1 = m_1 n_1, \qquad p + 1 = m_2 n_2, \qquad \gcd(m_1, m_2) = 2, \tag{23}$$

every prime factor of $m_1 m_2$ is at most y, and every prime factor of $n_1 n_2$ is greater than y. In particular, $\gcd(m_1, n_1) = \gcd(m_2, n_2) = 1$.

   We claim that

$$\frac{\varphi(m_1)}{m_1} \neq \frac{\varphi(m_2)}{m_2} \tag{24}$$

for sufficiently large x. In light of (6), it follows that $P(n)$ is the largest prime factor of the denominator of $\varphi(n)/n$. If $\varphi(m_1)/m_1 = \varphi(m_2)/m_2$, then $P(m_1) = P(m_2) = 2$ since $\gcd(m_1, m_2) = 2$. Thus, m_1 and m_2 are powers of 2 and

$$1 = \omega(m_1) + \omega(m_2) - 1$$

$$= \omega(m_1 m_2) = \omega(D_y(p^2 - 1))$$

$$= \omega_y(p^2 - 1) \in \left[1.5 \log \log y, 2.5 \log \log y\right]$$

because $p \notin \mathscr{E}_3(x)$. This is a contradiction for $x \geqslant 10^{483}$.

For convenient $p \leqslant x$, we have

$$S(p) = \frac{\varphi(p-1)}{p-1} - \frac{\varphi(p+1)}{p+1} = \frac{\varphi(m_1)\varphi(n_1)}{m_1 n_1} - \frac{\varphi(m_2)\varphi(n_2)}{m_2 n_2}.$$

We note that $S(p) \neq 0$ because otherwise $P(p-1) = P(p+1)$ by (6). Since $\gcd(p-1, p+1) = 2$, it would follow that $p-1$ and $p+1$ are powers of 2, which occurs only for $p = 3$. Lemma 2 ensures that $S(p)$ has the same sign as $\varphi(p-1) - \varphi(p+1)$ on a set of full density in the primes.

Since $p \notin \mathscr{E}_2(x)$, for large x we may use the inequality

$$|t + \log(1 - t)| \leqslant |t|^2, \qquad \text{for } |t| \leqslant \tfrac{1}{2},$$

to obtain

$$\frac{\varphi(n_1)}{n_1} = \prod_{\substack{r|(p-1)\\r>y}} \left(1 - \frac{1}{r}\right) = \exp\left(\sum_{\substack{r|(p-1)\\r>y}} \log\left(1 - \frac{1}{r}\right)\right)$$

$$= \exp\left(-\sum_{\substack{r|(p-1)\\r>y}} \frac{1}{r} + O\left(\left(\sum_{\substack{r|(p-1)\\r>y}} \frac{1}{r}\right)^2\right)\right)$$

$$= \exp\left(-h_y(p-1) + O\left(h_y(p-1)^2\right)\right)$$

$$= 1 + O\left(\frac{1}{y\sqrt{\log y}}\right),$$

in which the implied constant in the preceding can be taken to be 2. A similar inequality holds if n_1 is replaced by n_2. Consequently,

$$S(p) = \frac{\varphi(m_1)}{m_1}\left(1 + O\left(\frac{1}{y\sqrt{\log y}}\right)\right) - \frac{\varphi(m_2)}{m_2}\left(1 + O\left(\frac{1}{y\sqrt{\log y}}\right)\right)$$

$$= \frac{\varphi(m_1)}{m_1} - \frac{\varphi(m_2)}{m_2} + O\left(\frac{1}{y\sqrt{\log y}}\right), \tag{25}$$

in which the implied constant can be taken to be 4.

2.9 Weird Primes

A convenient prime $p \leqslant x$ is *weird* if

$$S(p)\left(\frac{\varphi(m_1)}{m_1} - \frac{\varphi(m_2)}{m_2}\right) < 0;$$

that is, if $S(p)$ and $\varphi(m_1)/m_1 - \varphi(m_2)/m_2$ have opposite signs [the second factor is nonzero if x is large; see (24)]. If this occurs, then (25) tells us that

$$\left| \frac{\varphi(m_1)}{m_1} - \frac{\varphi(m_2)}{m_2} \right| < \frac{4}{y\sqrt{\log y}}. \tag{26}$$

In general, one expects the sign of $S(p)$ to be determined by small primes; that is, those primes at most y. If p is weird, then the primes $q > y$ that divide $p^2 - 1$ conspire to overthrow the contribution of the small primes.

We say that a pair (m_1, m_2) of positive integers is *weird* if

$$24|m_1m_2, \quad m_1m_2|L_y, \quad \gcd(m_1, m_2) = 2, \quad \text{and} \quad (26) \text{ holds.}$$

What is the reason for the appearance of the number 24 in the preceding? If $p \geqslant 5$, then considering $p^2 - 1$ modulo 3 and 8 reveals that $24|(p^2 - 1)$. If $x \geqslant 10^9$, then $y \geqslant 3$ and hence $P(24) = 3$ and $24|D_y(p^2 - 1)$. Consequently, if we are searching for primes p for which $D_y(p^2 - 1) = m_1m_2$, it makes sense for us to insist that m_1m_2 is divisible by 24.

Lemma 4 *Let $y \geqslant \exp(48^6)$ and $D|L_y$ be a multiple of* 24.

(a) *The number of pairs (m_1, m_2) with $D = m_1m_2$ and $\gcd(m_1, m_2) = 2$ is $2^{\omega(D)}$.*
(b) *If $\omega(D) \in [1.5 \log\log y, 2.5 \log\log y]$ and $D/\varphi(D) \leqslant (\log y)^{1/3}$, then the number of weird pairs (m_1, m_2) with $D = m_1m_2$ is*

$$\ll \frac{2^{\omega(D)}}{\sqrt{\log\log y}}.$$

Proof

(a) In what follows, ν_p denotes the p-adic valuation function. Write

$$D = \prod_{q \in S} q^{\nu_q(D)},$$

in which S is a set of primes that contains 2, $\#S = \omega(D)$, and $\nu(2) \geqslant 3$. Since $\gcd(m_1, m_2) = 2$, it follows that

$$\nu_q(m_1) = \begin{cases} 1 \text{ or } \nu_2(D) - 1 & \text{if } q = 2, \\ 0 \text{ or } \nu_q(D) & \text{if } q \geqslant 3. \end{cases}$$

For each of the $\omega(D)$ primes in S, there are two possible choices for $\nu_q(m_1)$. Consequently, there are $2^{\omega(D)}$ possible pairs (m_1, m_2).

(b) Let

$$(m_1, m_2) = (2n_1, 2^{\nu_2(D)-1}n_2) \quad \text{or} \quad (2^{\nu_2(D)-1}n_1, 2n_2)$$

and

$$(m_1', m_2') = (2n_1', 2^{v_2(D)-1} n_2') \quad \text{or} \quad (2^{v_2(D)-1} n_1', 2n_2')$$

be weird, where n_1, n_2, n_1', n_2' are odd, and let $D = m_1 m_2 = m_1' m_2'$. Suppose toward a contradiction that $n_1 | n_1'$ and $n_1 < n_1'$. Then (26) says that

$$
\begin{aligned}
\left(\frac{\varphi(n_1)}{n_1}\right)^2 &= \frac{4\varphi(m_1)^2}{m_1^2} = \frac{4\varphi(m_1)}{m_1}\left(\frac{\varphi(m_2)}{m_2} + O\left(\frac{1}{y\sqrt{\log y}}\right)\right) \\
&= \frac{4\varphi(m_1)\varphi(m_2)}{m_1 m_2} + O\left(\frac{1}{y\sqrt{\log y}}\right) \quad (27) \\
&= \frac{2\varphi(m_1 m_2)}{m_1 m_2} + O\left(\frac{1}{y\sqrt{\log y}}\right) \\
&= \frac{2\varphi(D)}{D} + O\left(\frac{1}{y\sqrt{\log y}}\right) \\
&= \frac{2\varphi(D)}{D}\left(1 + O\left(\frac{1}{y(\log y)^{1/6}}\right)\right) \quad (28)
\end{aligned}
$$

since $D/\varphi(D) \leqslant (\log y)^{1/3}$. The implied constant in (28) is 16, in light of (26) and the absorption of $4\varphi(m_1)/m_1$ in (27). Similar reasoning yields an analogous expression for $\varphi(n_1')/n_1'$, with the same implied constant.

Let r be the smallest prime divisor of n_1'/n_1. Use the inequality

$$\frac{1+s}{1+t} \leqslant 1 + \frac{3}{2}|s - t| \leqslant 1 + \frac{3}{2}(|s| + |t|), \qquad |s|, |t| \leqslant \frac{1}{3},$$

and the fact that (28) holds for n_1 and n_1' to deduce that

$$
\begin{aligned}
1 + \frac{1}{r} < 1 + \frac{2}{r-1} &\leqslant \left(1 + \frac{1}{r-1}\right)^2 \\
&= \frac{1}{(1 - 1/r)^2} = \left(\frac{n_1'}{\varphi(n_1')} \cdot \frac{\varphi(n_1)}{n_1}\right)^2 \\
&\leqslant 1 + \frac{\frac{3}{2}(16 + 16)}{y(\log y)^{1/6}} = 1 + \frac{48}{y(\log y)^{1/16}}.
\end{aligned}
$$

Since $r | D$ and $D | L_y$, we have $r \leqslant y$ and hence

$$\frac{y(\log y)^{1/6}}{48} < r \leqslant y.$$

This is a contradiction if $y \geqslant \exp(48^6)$.

Hence, the set of odd components n_1 of the parts m_1 as (m_1, m_2) ranges over weird pairs has the property that no two divide each other. Identifying n_1 with the set of its odd prime factors, no two n_1 and n_1', as subsets, are contained one in another. Sperner's theorem[1] from combinatorics and Stirling's formula ensure that the number of such n_1, and hence the number of such pairs (m_1, m_2), is

$$\leq \binom{\omega(D)}{\lfloor \frac{\omega(D)}{2} \rfloor} \ll \frac{2^{\omega(D)}}{\sqrt{\omega(D)}} \ll \frac{2^{\omega(D)}}{\sqrt{\log \log y}}. \qquad \square$$

2.10 Conclusion

We have shown that the number of inconvenient primes at most x is $o(\pi(x))$ and hence they can be safely ignored. Each convenient prime p gives rise to a pair (m_1, m_2) as in (22).

Suppose that x is large. Let D be a multiple of 24 with

$$D | L_y, \quad \frac{D}{\varphi(D)} \leq (\log y)^{1/3}, \quad \text{and} \quad \omega(D) \in [1.5 \log \log y, 2.5 \log \log y].$$
$$(29)$$

We wish to count the primes $p \leq x$ for which $D_y(p^2 - 1) = D$. Denote this number by $\pi_D(x)$. To complete the proof of Theorem 1 in the case $\ell = 1$, it suffices to show that $(1/2 + o(1))\pi_D(x)$ of primes at most x have $S(p) > 0$ and $(1/2 + o(1))\pi_D(x)$ have $S(p) < 0$, and that the implied constant is uniform for all D as above.

Choose a pair (m_1, m_2) such that $D = m_1 m_2$ and $\gcd(m_1, m_2) = 2$. We want to count the primes $p \leq x$ such that $m_1 | (p - 1)$ and $m_2 | (p + 1)$; that is, such that

$$p \equiv 1 \ (\text{mod } m_1) \quad \text{and} \quad p \equiv -1 \ (\text{mod } m_2). \qquad (30)$$

Apply the Chinese Remainder Theorem with moduli $\frac{1}{2}m_1, m_2$ or $m_1, \frac{1}{2}m_2$, depending upon which of m_1, m_2 is exactly divisible by 2, to see that p belongs to an arithmetic progression a_{m_1, m_2} (mod $D/2$), with $\gcd(a_{m_1, m_2}, D/2) = 1$. However, we also want

$$\gcd\left(\frac{p^2 - 1}{m_1 m_2}, L_y\right) = 1.$$

For this, we need to work modulo

$$M_D := (D/2) \prod_{r \leq y} r.$$

To ensure that (30) holds, we do the following:

---

[1] A collection of sets that does not contain X and Y for which $X \subsetneq Y$ is a *Sperner family*. If S is a Sperner family whose union has a total of n elements, then $\#S \leq \binom{n}{\lfloor \frac{n}{2} \rfloor}$ [11].

- If $r \mid D$, then we want

$$p \equiv \varepsilon + r^{v_r(D/2)}\lambda \pmod{r^{v_r(D/2)+1}}$$

  for some $\lambda \in \{1, 2, \ldots, r-1\}$. Here, $\varepsilon = \pm 1$ according to whether $r^{v_r(D/2)}$ divides m_1 or m_2, respectively. For each $r \mid D$, there are $r-1$ possibilities for λ and hence there are $r-1$ possibilities for p modulo $r^{v_r(D/2)+1}$.

- If $r \nmid D$, then we want $p \equiv \lambda \pmod{r}$ for some $\lambda \notin \{0, 1, r-1\}$. For each $r \nmid D$, there are $r-3$ possibilities for p modulo r.

Thus, the number of progressions modulo M_D that can contain a prime p for which (30) occurs is

$$\prod_{r \mid \frac{D}{2}} (r-1) \prod_{\substack{r \leqslant y \\ r \nmid D}} (r-3). \tag{31}$$

By (10), the common modulus of all these progressions satisfies

$$M_D \leqslant L_y^2 \leqslant (\log x)^4$$

for large x. The Siegel–Walfisz theorem says that the number of primes in each progression is asymptotically

$$\frac{\pi(x)}{\varphi(M_D)} + O\left(xe^{-C\sqrt{\log x}}\right) \tag{32}$$

for some $C > 0$. Summing up over the number of progressions [or, more precisely, multiplying the (32) by the number of acceptable progressions (31)], and using the fact that $\varphi(M_D) = (D/2)\prod_{r \leqslant y}(r-1)$, we get a count of

$$\frac{2\pi(x)}{D} \prod_{\substack{r \leqslant y \\ r \nmid D}} \left(\frac{r-3}{r-1}\right) + O\left(xL_y^2 e^{-C\sqrt{\log x}}\right).$$

The count depends on D but not on the pair of divisors (m_1, m_2) of D. We now apply Lemma 4 and obtain

$$\pi_D(x) = \frac{2^{\omega(D)+1}\pi(x)}{D} \prod_{\substack{r \leqslant y \\ r \nmid D}} \left(\frac{r-3}{r-1}\right) + O(2^{\omega(D)}xL_y^2 e^{-C\sqrt{\log x}}). \tag{33}$$

Although it is not crucial to our proof, we show in Sect. 2.11 that

$$\sum_D \pi_D(x) = \pi(x)\left(1 + O\left(\frac{1}{\sqrt{\log\log y}}\right)\right), \tag{34}$$

where the index D runs over all D for which (29) holds, because the computation is of independent interest.

The product in (33) is less than 1 and bounded below by

$$\prod_{\substack{r\leqslant y \\ r\nmid D}}\left(\frac{r-3}{r-1}\right) \gg \prod_{r\leqslant y}\left(\frac{r-3}{r-1}\right) = \prod_{r\leqslant y}\left(\frac{r^2(r-3)}{(r-1)^3}\right)\left(1-\frac{1}{r}\right)^2$$

$$= \prod_{r\leqslant y}\left(1-\frac{3r-1}{(r-1)^3}\right)\prod_{r\leqslant y}\left(1-\frac{1}{r}\right)^2$$

$$\gg \prod_{r\leqslant y}\left(1-\frac{1}{r}\right)^2 \gg (\log y)^{-2} \tag{35}$$

by Mertens' asymptotic formula [10, §VII.29.1b, p. 259]. Since

$$2^{\omega(D)} \geqslant 2^{1.5\log\log y} \gg (\log y),$$

we examine the main term in (33) and conclude that

$$\pi_D(x) \gg \frac{\pi(x)}{D(\log y)^2}.$$

On the other hand, the error term in (33) is

$$O\left(2^{\omega(D)}xL_y^2 e^{-C\sqrt{\log x}}\right) = O\left(2^{2.5\log\log y}x(\log x)^4 e^{-C\sqrt{\log x}}\right)$$

$$= O\left((\log x)^4(\log y)^2 x e^{-C\sqrt{\log x}}\right)$$

$$= o\left(\frac{\pi(x)}{D(\log y)^3}\right) = o\left(\frac{\pi_D(x)}{\log y}\right).$$

There is a symmetry between non-weird pairs (m_1, m_2) with

$$\frac{\varphi(m_1)}{m_1} - \frac{\varphi(m_2)}{m_2} > 0 \qquad \text{and those with} \qquad \frac{\varphi(m_1)}{m_1} - \frac{\varphi(m_2)}{m_2} < 0$$

given by the transposition $(m_1, m_2) \mapsto (m_2, m_1)$. Indeed, we could return to (23) and insist that $m_1|(p+1)$ and $m_2|(p-1)$ instead. The subsequent asymptotic estimates go through in exactly the same manner. Via this transposition, we obtain an asymptotically equal count between the convenient primes $p \leqslant x$ corresponding to (m_1, m_2) and the convenient primes $p \leqslant x$ corresponding to (m_2, m_1). If only non-weird pairs (m_1, m_2) are taken into account, for fixed $D_y(p^2 - 1) = D$ this symmetry gives an asymptotically equal count of convenient primes $p \leqslant x$ with $S(p) > 0$ and with $S(p) < 0$.

Lemma 4 ensures that the number of weird pairs (m_1, m_2) with $D = m_1 m_2$ is $\ll 2^{\omega(D)}/\sqrt{\log \log y}$. As $x \to \infty$, we see from (33) that the number of primes at most x that arise from some weird pair is

$$\ll \frac{2^{\omega(D)} \pi(x)}{D\sqrt{\log \log y}} \prod_{\substack{r \leqslant y \\ r \nmid D}} \left(\frac{r-3}{r-1}\right) = O\left(\frac{\pi_D(x)}{\sqrt{\log \log y}}\right) = o(\pi_D(x)).$$

Recall that the non-weird, convenient primes have full density in the set of primes. Of such primes $p \leqslant x$, the argument above shows that an asymptotically equal amount have $S(p) > 0$ versus $S(p) < 0$ (recall that $S(p) = 0$ only for $p = 3$). This completes the proof of Theorem 1 in the case $\ell = 1$. $\square$

2.11   Sanity Check

Before extending the preceding proof to the case $\ell \geqslant 2$, it is helpful to perform a quick sanity check. Our goal here is to prove (34). In light of (33), it suffices to prove that

$$\sum_D \frac{2^{\omega(D)+1}}{D} \prod_{\substack{r \leqslant y \\ r \nmid D}} \left(\frac{r-3}{r-1}\right) = 1 + o(1), \tag{36}$$

in which the index D runs over all D for which (29) holds. In particular, (34) holds and the preceding product does not run over $r = 2, 3$. These developments seem remarkably fortuitous. Let us provide an independent derivation of (36), which will help corroborate some of the fine details in the preceding proof.

First write $D = 2^k D_1$, in which $k \geqslant 3$, and sum to obtain

$$\left(\sum_{k \geqslant 3} \frac{2^2}{2^k}\right)\left(\sum_{D_1 \text{ odd}} \frac{2^{\omega(D_1)}}{D_1}\right) = \sum_{D_1 \text{ odd}} \frac{2^{\omega(D_1)}}{D_1}.$$

Now write $D_1 = 3^k D_2$ and sum over $k \geqslant 1$ getting

$$\left(\sum_{k \geqslant 1} \frac{2}{3^k}\right)\left(\sum_{\gcd(D_2,6)=1} \frac{2^{\omega(D_2)}}{D_2}\right) = \sum_{\gcd(D_2,6)=1} \frac{2^{\omega(D_2)}}{D_2}.$$

For the rest, we use multiplicativity to say that the sum in (36) is

$$\prod_{r \leqslant y}\left(\frac{r-3}{r-1}\right) \prod_{5 \leqslant r \leqslant y}\left(1+\frac{2}{(r-3)/(r-1)}\sum_{k \geqslant 1}\frac{1}{r^k}\right).$$

However, this is not strictly correct since the sum above stops at the largest power $r^b \leqslant y$. Moreover, the sum runs over all D without restrictions such as $\omega(D) \in [1.5 \log \log y, 2.5 \log \log y]$ or $D/\varphi(D) \leqslant (\log y)^{1/3}$. We deal with these omissions shortly. For the time being, let us ignore these restrictions. Then the amount inside the Euler factor is

$$1+\frac{2(r-1)}{r-3}\frac{1}{r-1}=1+\frac{2}{r-3}=\frac{r-1}{r-3},$$

which cancels with the outside $(r-3)/(r-1)$.

Now we must examine the errors. There are essentially four types:

(a) In each Euler factor we only sum up to r^b, in which b is maximal such that $r^b \leqslant y$. By extending the sum to infinity we incurred an error of

$$\frac{2}{(r-3)/(r-1)}\sum_{k \geqslant b+1}\frac{1}{r^k}=\frac{2}{r^b(r-3)}=O\left(\frac{1}{y}\right).$$

For $5 \leqslant r \leqslant y$, the actual Euler factor is

$$\frac{r-1}{r-3}+O\left(\frac{1}{(r-3)y}\right)=\frac{r-1}{r-3}\left(1+O\left(\frac{1}{y}\right)\right)$$

Similar considerations apply for $r=2,3$. The total multiplicative error is

$$\left(1+O\left(\frac{1}{y}\right)\right)^{\pi(y)}=1+O\left(\frac{\pi(y)}{y}\right)=1+O\left(\frac{1}{\log y}\right).$$

(b) We consider only D such that $D/\varphi(D) \leqslant (\log y)^{1/3}$. Let the set of remaining D be denoted $\mathscr{D}$. For $D \in \mathscr{D}$, we have

$$\frac{1}{D} \leqslant \frac{1}{(\log y)^{1/3}}\cdot\frac{1}{\varphi(D)}.$$

Applying this inequality and extending then the sum over all possible D, the piece of the sum over $\mathscr{D}$ is at most

$$\ll \frac{\pi(x)}{(\log y)^{1/3}}\sum_{D}\frac{2^{\omega(D)+1}}{\varphi(D)}\prod_{\substack{r \leqslant y \\ r \nmid D}}\left(\frac{r-3}{r-1}\right).$$

We separate out the power of 2 in D as $D = 2^k D_1$ with $k \geqslant 3$ getting an Euler factor corresponding to 2 of

$$4 \left(\frac{1}{\varphi(8)} + \frac{1}{\varphi(16)} + \cdots + \frac{1}{\varphi(2^k)} + \cdots \right) = 2.$$

Then we separate out a factor of 3 from D_1 writing it as $D_1 = 3^k D_2$, getting an Euler factor corresponding to 3 of

$$2 \left(\frac{1}{\varphi(3)} + \frac{1}{\varphi(9)} + \cdots + \frac{1}{\varphi(3^k)} + \cdots \right) = \frac{3}{2}.$$

For the remaining primes $r \geqslant 5$, we form the Euler product getting

$$\frac{\pi(x)}{(\log y)^{2/3}} \prod_{5 \leqslant r \leqslant y} \left(\frac{r-3}{r-1} \right) \prod_{r=5}^{y} \left(1 + \frac{2}{(r-3)/(r-1)} \sum_{b \geqslant 1} \frac{1}{\varphi(r^b)} \right).$$

The factor inside the parentheses is

$$1 + \frac{2}{(r-3)/(r-1)} \frac{1}{(r-1)(1-1/r)} = 1 + \frac{2r}{(r-1)(r-3)}$$

$$= \frac{r^2 - 2r + 3}{(r-1)(r-3)}$$

$$= \frac{(r-1)}{(r-3)} \left(1 + \frac{2}{(r-1)^2} \right).$$

Multiply this by the outside factor $(r-3)/(r-1)$ and get

$$\left(\frac{r-3}{r-1} \right) \left(1 + \frac{2}{(r-3)/(r-1)} \sum_{b \geqslant 1} \frac{1}{\varphi(r^b)} \right) = 1 + \frac{2}{(r-1)^2}.$$

Taking the product of the factors above over $r \in [5, y]$, we get a convergent product. Consequently,

$$\pi(x) \sum_{\substack{D \mid L_y, 24 \mid D \\ D/\varphi(D) > (\log y)^{1/3}}} \frac{2^{\omega(D)+1}}{D} \prod_{\substack{5 \leqslant \leqslant y \\ r \nmid D}} \left(\frac{r-3}{r-1} \right) \ll \frac{\pi(x)}{(\log y)^{1/3}}.$$

(c) We need to consider D with $\omega(D) \notin [1.5 \log \log y, 2.5 \log \log y]$. From the preceding material and (35), we have

$$\prod_{5 \leqslant r \leqslant y} \left(1 + \sum_{b \geqslant 1} \frac{2}{(r-3)/(r-1)} \sum_{b \geqslant 1} \frac{1}{r^b} \right) \asymp \left(\prod_{5 \leqslant r \leqslant y} \left(\frac{r-3}{r-1} \right) \right)^{-1} \asymp (\log y)^2.$$

We first deal with D with many prime factors. Consider the multiplicative function defined for prime powers r^b with $r \geqslant 5$ by

$$f(r) = \frac{2(r-1)}{(r-3)r^b}.$$

If $K := \lfloor 2.5 \log \log y \rfloor$, then

$$\sum_{\substack{r | D_2 \Longrightarrow r \in [5,y] \\ \omega(D_2) > K-2}} f(D_2) \quad \leqslant \quad \sum_{k > K-2} \frac{1}{k!} \left(\sum_{\substack{r \in [5,y] \\ b \geqslant 1}} f(r^b) \right)^k.$$

We have

$$S = \sum_{\substack{r \in [5,y] \\ b \geqslant 1}} f(r^b) = \sum_{\substack{b \geqslant 1 \\ 5 \leqslant r \leqslant y}} \left(\frac{2}{r^b} + \frac{2}{r^b} \left(\frac{r-1}{r-3} - 1 \right) \right)$$

$$= \sum_{\substack{r \in [2,y] \\ b \geqslant 1}} \frac{2}{r^b} + O \left(\sum_{\substack{r \geqslant 5 \\ b \geqslant 1}} \frac{1}{r^{b+1}} \right)$$

$$= 2 \log \log y + O(1),$$

in which we have used Mertens' theorem [10, §VII.28.1b]. In the sum $\sum_{k>K} S^k/k!$, the ratio of two consecutive terms is

$$\frac{S^{k+1}/(k+1)!}{S^k/k!} = \frac{S}{k+1} = \frac{2 \log \log y + O(1)}{2.5 \log \log y + O(1)} < \frac{5}{6}$$

for $k > K-2$ and large x, so the first term dominates. With $K! > (K/e)^K$, the contribution of D with $\omega(D) > 2.5 \log \log y$ is at most

$$\ll \frac{S^{K-2}}{(K-2)!} \leqslant \left(\frac{2e \log \log y + O(1)}{2.5 \log \log y + O(1)} \right)^{2.5 \log \log y + O(1)} \ll (\log y)^c,$$

in which $c = 2.5 \log(2e/2.5) < 1.95$. Multiplying this by (see (35))

$$\prod_{5 \leqslant r \leqslant y} \left(\frac{r-3}{r-1} \right) \ll (\log y)^{-2},$$

we obtain

$$
\pi(x) \sum_{\substack{D|L_y \\ 24|D \\ \omega(D)>2.5\log\log y}} \frac{2^{\omega(D)+1}}{D} \prod_{\substack{5\leqslant r\leqslant y \\ r\nmid D}} \left(\frac{r-3}{r-1}\right) \ll \frac{\pi(x)}{(\log y)^{0.5}}.
$$

We use a similar argument for D with $\omega(D) < 1.5\log\log y$. In this case, let $K_1 := \lfloor 1.5\log\log y \rfloor$. We have to deal with

$$
\sum_{\substack{D_2:r|D_2 \Longrightarrow r\in[5,y] \\ \omega(D_2)<1.5\log\log y-2}} f(D_2) \leqslant \sum_{k\leqslant K_1-2} \frac{1}{k!}S^k.
$$

For $k > K_1 - 2$ and large x, the ratio of any two consecutive terms above is

$$
\frac{S^{k+1}/(k+1)!}{S^k/k!} = \frac{S}{k+1} > \frac{2\log\log y + O(1)}{1.5\log\log y + O(1)} > \frac{5}{4},
$$

it follows that the last term dominates. Thus, this sum is at most

$$
\left(\frac{2e\log\log y + O(1)}{K_1 - 2}\right)^{K_1-2} = \left(\frac{2e\log\log y + O(1)}{1.5\log\log y + O(1)}\right)^{1.5\log\log y+O(1)}
$$
$$
\ll (\log y)^{c_1},
$$

in which where $c_1 = 1.5\log(2e/1.5) < 1.95$. Consequently, the contribution of D with $\omega(D) < 1.5\log\log y$ to the sum defining $\pi_D(x)$ is

$$
\ll \frac{\pi_D(x)}{(\log y)^{0.05}}.
$$

Putting everything together we obtain (36), which is equivalent to (34).

3 Proof of Theorem 1 for $\ell \geqslant 2$

The proof for $\ell \geqslant 2$ follows largely on the lines of the case $\ell = 1$, although there are a number of minor adjustments that must be made. For example, in Lemma 4 we assumed that D is a multiple of 24. Elementary considerations reveal that the following adjustments are necessary for various values of ℓ:

(a) D is coprime to all primes that divide ℓ.
(b) D is odd if ℓ is even.
(c) D is a multiple of 8 if ℓ is odd.

(d) D is a multiple of 3 if and only if ℓ is not.

More significant modifications are discussed below.

3.1 The Case of Equality

We need a variant of the inequality (3). The estimate provided by the following theorem involves two special cases. Numerical evidence strongly suggests that this distinction is not simply a by-product of our proof; see Table 1. If we replace the use of the Brun sieve in what follows with an appeal to the Bateman–Horn conjecture [1], then the larger of the two upper bounds becomes an asymptotic equivalence if the appropriate constant factor is introduced.

Theorem 2 *For each $\ell \geqslant 1$,*

$$\#\{p \leqslant x : \varphi(p - \ell) = \varphi(p + \ell)\} \ll \begin{cases} \dfrac{x}{(\log x)^3} & \text{if } \ell = 4^n - 1, \\[2ex] \dfrac{x}{e^{(\log x)^{1/3}}} & \text{otherwise.} \end{cases}$$

Proof In Lemma 1, let $k = 2\ell$ and $n = p - \ell$, in which p is prime. Suppose that j and $j + 2\ell$ have the same prime factors. Since

$$p = j\left(\frac{(j + 2\ell)t}{g} + 1\right) + \ell, \tag{37}$$

it follows that j is not divisible by any prime factor of ℓ and hence $g = \gcd(j, 2\ell) = 2$. Thus, $j = 2^m$ and $j + 2\ell = 2^{m+n}$ for some $m, n \geqslant 1$. Then $2^{m+n} = 2^m + 2\ell$ and

$$\ell = 2^{m-1}(2^n - 1).$$

If $m \geqslant 2$, then ℓ is even and (37) implies that $2 \mid p$, a contradiction. Thus, the upper bound from Lemma 1 applies in this case.
    If $m = 1$, then $j = 2$ and $\ell = 2^n - 1$. Then

$$p = 2(\ell + 1)t + (\ell + 2), \qquad q = t + 1 \qquad \text{and} \qquad r = (\ell + 1)t + 1$$

and we count $t \leqslant \frac{x - (\ell + 2)}{2(\ell + 1)} \sim \frac{x}{2\ell}$ for which p, q, r are simultaneously prime. Let

$$f_1(t) = 2^{n+1}t + (2^n + 1), \qquad f_2(t) = t + 1, \qquad \text{and} \quad f_3(t) = 2^n t + 1.$$

If n is odd, then

$$f_1(0) \equiv f_2(2) \equiv f_3(1) \equiv 0 \,(\text{mod } 3).$$

There are three possibilities:

(a) If $f_1(0) = 3$, then $f_2(0) = 1$ is not prime and no prime triples are produced.
(b) If $f_2(2) = 3$, then for each odd n, at most one prime triple is produced.[2]
(c) If $f_3(1) = 3$, then $n = 1$ and only the prime triple $(7, 2, 3)$ is produced.

In each case, the upper bound from Lemma 1 dominates.

If n is even, then none of f_1, f_2, f_3 vanish identically modulo any prime. The Brun sieve says that the number of $p \leqslant x$ for which p, q, r are prime is $O(x/(\log x)^3)$, which dominates the estimate from Lemma 1.

3.2 A More General Comparison Lemma

The next adjustment that is required is an analogue of the comparison lemma (Lemma 2). This turns out to be more involved than expected. In fact, we first need a generalization of the "Turán–Kubilius"-type result from Lemma 3. Since this is a minor variant of an existing result, we only sketch the proof.

Lemma 5 *For each $\ell \geqslant 1$,* $\displaystyle\sum_{p \leqslant x} (\omega_y(p \pm \ell) - \log\log y)^2 = O(\pi(x)\log\log y)$.

Proof Since $y \leqslant (\log x)^2$, apply the Siegel–Walfisz theorem to obtain

$$\sum_{p \leqslant x} \omega_y(p \pm \ell) = \pi(x)\log\log y + O(\pi(x)),$$

$$\sum_{p \leqslant x} \omega_y(p \pm \ell)^2 = \pi(x)(\log\log y)^2 + O(\pi(x)\log\log y),$$

in which the $\log\log y$ term arises from an application of Mertens' theorem [10, §VII.28.1b]. Now expand $\sum_{p \leqslant x} (\omega_y(p \pm \ell) - \log\log y)^2$ and apply the preceding. ∎

The direct generalization of Lemma 2 for $\ell \geqslant 2$ runs into trouble. If ℓ is sufficiently large, then the $+1$ in (8) becomes too large for the same argument to work. The evenness of $\varphi(p - \ell) - \varphi(p + \ell)$ is no longer sufficient to push the argument through. Fortunately, we are able to employ the following lemma instead.

Lemma 6 *For $\ell, m \geqslant 1$,*

$$\#\{p \leqslant x : \varphi(p \pm \ell) \equiv 0 \,(\mathrm{mod}\, 2^m)\} \sim \pi(x).$$

---

[2]The only odd $n < 99$ for which a prime triple arises in this manner are $n = 1, 3, 7, 15$, from which we obtain the triples $(11, 3, 5)$, $(41, 3, 17)$, $(641, 3, 257)$, and $(163{,}841, 3, 65{,}537)$.

Proof Fix $\ell \geqslant 1$. Let $\omega(n)$ denote the number of distinct prime divisors of n. Then $2^{\omega(n)-1} | \varphi(n)$ since $\varphi(n) = \prod_{p^a \| n} p^{a-1}(p-1)$. If

$$2 + \log_2 \ell \leqslant m + 1 \leqslant \omega(p \pm \ell), \tag{38}$$

then

$$2\ell \leqslant 2^m \leqslant 2^{\omega(p \pm \ell)-1} \quad \text{and hence} \quad \varphi(p \pm \ell) \equiv 0 \pmod{2^m}.$$

Thus, it suffices to show that the set of primes $p \leqslant x$ for which (38) fails has a counting function that is $o(\pi(x))$. Let x be so large that $y = \log \log x$ satisfies

$$2 + \log_2 \ell \leqslant \tfrac{1}{2} \log \log y$$

and let

$$\mathcal{E}(x) = \left\{ p \leqslant x : \omega(p - \ell) < 2 + \log_2 \ell \quad \text{or} \quad \omega(p + \ell) < 2 + \log_2 \ell \right\}.$$

Let $\omega_y(n)$ denote the number of distinct prime factors $q \leqslant y$ of n. If $p \in \mathcal{E}(x)$, then

$$\omega_y(p - \ell) < \tfrac{1}{2} \log \log y \quad \text{or} \quad \omega_y(p + \ell) < \tfrac{1}{2} \log \log y,$$

and hence

$$\omega_y(p - \ell) \notin [\tfrac{1}{2} \log \log y, \tfrac{3}{2} \log \log y] \quad \text{or} \quad \omega_y(p + \ell) \notin [\tfrac{1}{2} \log \log y, \tfrac{3}{2} \log \log y].$$

Then

$$\tfrac{1}{4}(\log \log y)^2 \leqslant (\omega_y(p-\ell) - \log \log y)^2 \quad \text{or} \quad \tfrac{1}{4}(\log \log y)^2 \leqslant (\omega_y(p+\ell) - \log \log y)^2.$$

Lemma 5 ensures that

$$\tfrac{1}{4}(\log \log y)^2 \# \mathcal{E}(x) \leqslant \sum_{p \in \mathcal{E}(x)} (\omega_y(p - \ell) - \log \log y)^2 + (\omega_y(p + \ell) - \log \log y)^2$$

$$\leqslant \sum_{p \leqslant x} (\omega_y(p - \ell) - \log \log y)^2 + (\omega_y(p + \ell) - \log \log y)^2$$

$$= O(\pi(x) \log \log y).$$

Thus,

$$\# \mathcal{E}(x) \ll \frac{\pi(x)}{\log \log y} = o(\pi(x)). \qquad \square$$

Our replacement for the comparison lemma is the following. Since the exceptional set is $o(\pi(x))$, it will not affect the proof of Theorem 1 in the case $\ell \geqslant 2$.

Lemma 7 *For each $\ell \geqslant 1$, the set of primes p for which $\varphi(p - \ell) - \varphi(p + \ell)$ and*

$$S(p) := \frac{\varphi(p - \ell)}{p - \ell} - \frac{\varphi(p + \ell)}{p + \ell} \tag{39}$$

have the same sign has counting function $\sim \pi(x)$.

Proof By Theorem 2, it suffices to show that the set of primes p for which

$$\varphi(p - \ell) > \varphi(p + \ell) \quad \Longleftrightarrow \quad \frac{\varphi(p - \ell)}{p - \ell} > \frac{\varphi(p + \ell)}{p + \ell} \tag{40}$$

has counting function $\sim \pi(x)$. The forward implication is straightforward, so we focus on the reverse. If the inequality on the right-hand side of (40) holds, then

$$0 < (p + \ell)\varphi(p - \ell) - (p - \ell)\varphi(p + \ell)$$

$$= p[\varphi(p - \ell) - \varphi(p + \ell)] + \ell\varphi(p - \ell) + \ell\varphi(p + \ell)$$

$$\leqslant p[\varphi(p - \ell) - \varphi(p + \ell)] + \ell(p - \ell) + \ell(p + \ell)$$

$$= p[\varphi(p - \ell) - \varphi(p + \ell) + 2\ell].$$

Let $2^m > 2\ell$ and apply lemma Lemma 6 to conclude that $\varphi(p - \ell) - \varphi(p + \ell) > 0$ for p in a set with counting function $\sim \pi(x)$.

3.3   Final Ingredient

The only other ingredient necessary to consider $\ell \geqslant 2$ is a replacement for the estimate (21). We include the proof for completeness.

Lemma 8 $\displaystyle\sum_{p \leqslant x} \frac{p \pm \ell}{\varphi(p \pm \ell)} \ll \pi(x).$

Proof Let $\sigma(n)$ denote the sum of the divisors of n. Then

$$\frac{\sigma(n)}{n} = \sum_{d \mid n} \frac{1}{d} \quad \text{and} \quad \frac{6}{\pi^2} < \frac{\sigma(n)\varphi(n)}{n^2}, \quad \text{for } n \geqslant 2;$$

see [10, §I.3.5]. The Siegel–Walfisz theorem provides $C > 0$ so that

$$\sum_{p \leqslant x} \frac{p \pm \ell}{\varphi(p \pm \ell)} \ll \sum_{p \leqslant x} \frac{\sigma(p \pm \ell)}{(p \pm \ell)} = \sum_{p \leqslant x} \sum_{d \mid (p \pm \ell)} \frac{1}{d}$$

$$= \sum_{d \leqslant x} \frac{1}{d} \sum_{\substack{p \leqslant x \\ p \equiv \mp 1 \,(\mathrm{mod}\, d)}} 1 = \sum_{d \leqslant x} \frac{\pi(x; \mp \ell, d)}{d}$$

$$= \sum_{d \leqslant (\log x)^3} \frac{\pi(x; \mp 1, d)}{d} + \sum_{(\log x)^3 \leqslant d \leqslant x} \frac{\pi(x; \mp 1, d)}{d}$$

$$\leqslant \sum_{d \leqslant (\log x)^3} \left(\frac{\pi(x)}{d\varphi(d)} + O\left(xe^{-C\sqrt{\log x}} \right) \right) + \sum_{(\log x)^3 \leqslant d \leqslant x} \frac{x}{d^2}$$

$$\leqslant \pi(x) \sum_{1 \leqslant d < \infty} \frac{1}{d\varphi(d)} + O\left(x(\log x)^3 e^{-C\sqrt{\log x}} \right)$$

$$+ x \sum_{(\log x)^3 \leqslant d < \infty} \frac{1}{d^2}$$

$$\ll \pi(x) + x(\log x)^3 e^{-C\sqrt{\log x}} + \frac{x}{(\log x)^3}$$

$$\ll \pi(x). \qquad \square$$

Acknowledgements We thank the referee for suggestions which improved the quality of our paper. SRG supported by NSF grant DMS-1265973, a David L. Hirsch III and Susan H. Hirsch Research Initiation Grant, and the Budapest Semesters in Mathematics (BSM) Director's Mathematician in Residence (DMiR) program. F. L. was supported in part by grants CPRR160325161141 and an A-rated researcher award both from the NRF of South Africa and by grant no. 17-02804S of the Czech Granting Agency. Part of this work was done while F. Luca visited the Max Planck Institute for Mathematics in Bonn, Germany in 2017.

References

1. P.T. Bateman, R.A. Horn, A heuristic asymptotic formula concerning the distribution of prime numbers. Math. Comput. **16**, 363–367 (1962). MR 0148632
2. P. Erdős, C. Pomerance, A. Sárközy, On locally repeated values of certain arithmetic functions. II. Acta Math. Hungar. **49**(1–2), 251–259 (1987). MR 869681
3. P. Erdős, On the normal number of prime factors of $p-1$ and some related problems concerning Euler's ϕ-function. Q. J. Math. **6**(1), 205–213 (1935)
4. S.R. Garcia, E. Kahoro, F. Luca, Primitive root discrepancy for twin primes. https://arxiv.org/abs/1705.02485

5. S.W. Graham, J.J. Holt, C. Pomerance, On the Solutions to $\varphi(n) = \varphi(n+k)$, in *Number Theory in Progress, Vol. 2 (Zakopane-Kościelisko, 1997)* (de Gruyter, Berlin, 1999), pp. 867–882. MR 1689549
6. H.N. Shapiro, Addition of functions in probabilistic number theory. Commun. Pure Appl. Math. **26**, 55–84 (1973)
7. F. Luca, C. Pomerance, On some problems of Mąkowski-Schinzel and Erdős concerning the arithmetical functions φ and σ. Colloq. Math. **92**(1), 111–130 (2002). MR 1899242
8. Y. Motohashi, On a property of $p - 1$. Proc. Jpn. Acad. **45**, 639–640 (1969). MR 0266880
9. K. Prachar, *Primzahlverteilung* (Springer, Berlin/Göttingen/Heidelberg, 1957). MR 0087685
10. J. Sándor, D.S. Mitrinović, B. Crstici, *Handbook of Number Theory. I* (Springer, Dordrecht, 2006). Second printing of the 1996 original. MR 2186914
11. E. Sperner, Ein Satz über Untermengen einer endlichen Menge. Math. Z. **27**(1), 544–548 (1928). MR 1544925
12. G. Tenenbaum, *Introduction to Analytic and Probabilistic Number Theory*. Graduate Studies in Mathematics, vol. 163, 3rd edn. (American Mathematical Society, Providence, RI, 2015). Translated from the 2008 French edition by Patrick D. F. Ion. MR 3363366
13. T. Yamada, On equations $\sigma(n) = \sigma(n + k)$ and $\phi(n) = \phi(n + k)$. J. Combinatorics Number Theory (to appear)

Vinogradov's Mean Value Theorem as an Ingredient in Polynomial Large Sieve Inequalities and Some Consequences

Karin Halupczok

Abstract We discuss the role of Vinogradov's mean value theorem in polynomial large sieve inequalities. We present an application to the distribution of fractions with k-th power denominators. Moreover, polynomial large sieve inequalities lead to a new approach of understanding certain aspects of prime distribution in arithmetic progressions, namely in Bombieri–Vinogradov's theorem with moduli of special multivariable polynomial shape. The recent progress leading to the main conjecture of Vinogradov's mean value theorem has an impact to such applications.

1 Introduction

The classical large sieve inequality is a central topic in sieve theory. It states that for any sequence (v_n) of complex numbers,

$$\sum_{q \leq Q} \sum_{\substack{a \bmod q \\ \gcd(a,q)=1}} \left| S\left(\frac{a}{q}\right) \right|^2 \leq (N + Q^2) \sum_{M < n \leq M+N} |v_n|^2, \tag{1}$$

where

$$S(\alpha) := \sum_{M < n \leq M+N} v_n e(\alpha n),$$

with integers $Q, M, N \geq 1$, and for real α, we let $e(\alpha) := \exp(2\pi i \alpha)$ denote the complex exponential function. Usually, we write $\|v\|^2$ for $\sum_{M < n \leq M+N} |v_n|^2$.

Here q is called modulus. The terminology is inherited from the widely known large sieve interpretation in which q stands for the moduli of residue classes that one sieves. The explanation that shows the connection to the large sieve can be found

K. Halupczok (✉)
Mathematisches Institut der Heinrich-Heine-Universität Düsseldorf, Düsseldorf, Germany
e-mail: Karin.Halupczok@uni-duesseldorf.de

© Springer International Publishing AG, part of Springer Nature 2018   97
J. Pintz, M. Th. Rassias (eds.), *Irregularities in the Distribution of Prime Numbers*,
https://doi.org/10.1007/978-3-319-92777-0_5

at many places in modern textbooks on sieve theory; a good survey is given by
H. Montgomery in [13], which is still recommendable. By now, the theory of the
large sieve is classical and one of the main tools of the number theorist.

Also many variants of the large sieve have been found to be useful in number
theory, especially those with moduli q coming from certain subsets of the full set of
integers up to a bound. In particular, the study of sparse subsets has been shown to
be successful.

As an example, the left-hand side in the large sieve inequality with q replaced by
polynomial values $P(q)$ can be considered to be natural in this context, so let

$$\Sigma_P := \sum_{q \leq Q} \sum_{\substack{a \bmod P(q) \\ \gcd(a, P(q))=1}} \left| S\left(\frac{a}{P(q)}\right) \right|^2.$$

In this article, we call an upper bound for Σ_P a polynomial LSI, where LSI
stands for large sieve inequality. Polynomials P in several variables may also be
considered, but the vast literature usually restricts to the already quite complicated
one-variable case.

Over the last years, a very different focus of interest in number theory was the
progress around the so-called Vinogradov mean value theorem (VMVT for short)
initiated by T. Wooley in a series of papers, see in particular [21]. The progress is
based on a method called efficient congruencing.

The present author showed in [9] that Weyl sum estimates incorporating VMVT
can be used as a tool to prove a new bound in the polynomial LSI with $P(x) = x^k$,
inspired by the work of S. Baier and L. Zhao in [1, 25].

In the meantime, further substantial steps towards the main conjecture in
Vinogradov's mean value theorem were made, and the progress culminated in the
spectacular proof of the main conjecture for all degrees exceeding 3 by J. Bourgain,
C. Demeter and L. Guth in [5]. Their proof relies heavily on the use of a method
called ℓ^2-decoupling. Together with Wooley's proof of the cubic case in [22], the
main conjecture in VMVT is solved now completely. A comprehensive overview of
the recent progress in VMVT is given by L. Pierce in [16]. Note that very recently,
Wooley gives a new complete proof of VMVT in the preprint [23] using a refined
version of his method called nested efficient congruencing.

By this remarkable work, the best possible exponent in VMVT has been
confirmed. So the proof in [9], with only few obvious changes to be made, also
leads to a new improvement in polynomial LSI bounds. As a consequence, in the
one-variable case, we are now able to state the following quite general result; it is
the best we can currently prove by known methods.

Theorem 1 ([9, Thm. 1.1], with [5, 22]) *For any monic real polynomial P of
degree $k \geq 2$ in one variable, we have*

$$\Sigma_P \ll_{\varepsilon,k} (Q^{k+1} + A_k(N, Q))(NQ)^{\varepsilon} \|v\|^2$$

with the expression

$$A_k(Q, N) := N Q^{1-1/k(k-1)} + N^{1-1/k(k-1)} Q^{1+1/(k-1)}.$$

In other words, the exponent δ from the article [9] can now be improved to $\delta = 1/k(k-1)$ by the factor 2 due to VMVT.

Wooley's breakthroughs around VMVT have been generalized by S.T. Parsell, S.M. Prendiville and T. Wooley in [15] to a multidimensional setting, i.e., to sums of multivariate monomials. A proof of the exact multidimensional analogue of the main conjecture has not been published yet.

In [10], the multivariable version [15, Thm. 10.1] for Weyl sums has already led to improvements in the multivariable setting of the polynomial LSI. Similar multivariable versions of such LSIs have not been found before.

In this article we give some current polynomial LSI bounds, some conjectural considerations and we mention the crucial link from VMVT to polynomial LSIs, namely Weyl sums estimates. We proceed giving an application of the one-variable polynomial LSI to the distribution of fractions a/q^k for $Q < q \leq 2Q$ that have not been published in print before.

Further, we discuss a consequence of the polynomial LSI for the distribution of primes in the style of Bombieri–Vinogradov's theorem with suitable polynomial moduli in many variables. We conclude with some thoughts on its applicability to the problem of bounded gaps between primes, in particular to the work of J. Maynard in [12].

A Note on Implicit Constants By k, ℓ we denote positive integers, usually k denotes a polynomial degree and ℓ the number of variables. Let ε be an arbitrary small positive real number that may change its specific value from one step to another. In this article, we usually suppress the dependence of the implicit constants on k, ℓ or ε in our notation and simply write $\ll$ for $\ll_{k,\ell,\varepsilon}$ or $\ll_{k,\varepsilon}$.

2 Overview of Current Polynomial LSI Bounds and Conjectures

Zhao proved in [25] the bound

$$\Sigma_{x^k} \ll \left(Q^{k+1} + N Q^{1-1/\kappa} + N^{1-1/\kappa} Q^{1+k/\kappa} \right) (NQ)^{\varepsilon} \|v\|^2, \tag{2}$$

where $\kappa := 2^{k-1}$. In [1], Baier and Zhao improved this bound and showed that

$$\Sigma_{x^k} \ll \left(Q^{k+1} + N + N^{1/2+\varepsilon} Q^k \right) (\log \log 10NQ)^{k+1} \|v\|^2. \tag{3}$$

This result is now superseded by Theorem 1 in the range $Q^k \ll N \ll Q^{2k-2+2/k(k-1)}$ when $k \geq 3$.

Based on averaging considerations, in [25], Zhao gave the following conjectural estimate for the general one-variable case.

Conjecture 1 For all real monic polynomials P of degree $k \geq 2$,

$$\Sigma_P \ll \left(Q^{k+1} + N\right) (NQ)^\varepsilon \|v\|^2.$$

But Theorem 1 is still quite far away from this conjecture.

In [2], Baier and Zhao proved the bound

$$\Sigma_{x^2} \ll \left(Q^3 + N + \min\left(NQ^{1/2}, N^{1/2}Q^2\right)\right)(NQ)^\varepsilon \|v\|^2 \tag{4}$$

for squares. This bound may be seen as a hint that the following milder conjecture than Conjecture 1 might be within reach of further refinements.

Conjecture 2 For all real monic polynomials P of degree $k \geq 2$,

$$\Sigma_P \ll \left(Q^{k+1} + N + \min\left(NQ^{1-1/k(k-1)}, N^{1-1/k(k-1)}Q^{1+1/(k-1)}\right)\right)(NQ)^\varepsilon \|v\|^2.$$

We observe that Conjecture 2 has already been confirmed by Baier and Zhao [2] in the case $k = 2$, but not yet in other degrees. Still, Conjecture 2 is far from Zhao's Conjecture 1, especially for larger k. In the meantime, one might believe that Zhao's conjecture could be too good to be true, and that Conjecture 2 should be the correct bound. It is clear that any improvements will have an impact to the exponents in the known applications.

For suitable polynomials P of degree k in ℓ variables, the multidimensional version in [10] yields the bound

$$\Sigma_P \ll \left(Q^{k+\ell} + Q^{\ell-\sigma'} + N^{1-\sigma'}Q^{\ell+k\sigma'}\right)(NQ)^\varepsilon \|v\|^2, \tag{5}$$

where

$$\sigma' = \sigma(k, \ell) = 1/2rk \text{ with } r = \binom{k+\ell-1}{\ell} - 1, \tag{6}$$

and σ' is the exact exponent coming from the multidimensional version in [15]. (A small technical correction needs to be made: In (5), the tuples $\mathbf{q}$ in the sum Σ_P run over tuples with coordinates in the interval $[Q, 2Q)$.)

Several instances of this bound (5) have been discussed in [10]. It turned out that these polynomial LSI bounds are superior to standard approaches only if the polynomial degree is bigger than the number of variables. So in some desirable multivariable applications, the bound gets too weak, but it is strong if the degree is much bigger compared to the number of variables.

3   The Path from VMVT to Polynomial Large Sieve Inequalities

The key to follow this path is the study of Weyl sums.

A Weyl sum is an exponential sum of the form $\sum_{q \leq Q} e(P(q))$ with a real polynomial P in one variable. A good overview of Weyl sums, especially of nontrivial estimates using VMVT can be found in [14] by Montgomery.

A multidimensional Weyl sum is a sum of the form $\sum_{\mathbf{q} \leq Q} e(P(\mathbf{q}))$ with a real polynomial P in ℓ many variables $X_1, \ldots, X_\ell$, so the sum runs over ℓ-tuples $\mathbf{q} \in \mathbb{N}^\ell$, where $\mathbf{q} \leq Q$ means that all coordinates of $\mathbf{q}$ are at most Q.

Let $k \geq 2$ be the degree of P and let σ' be the exponent from (6).

Fix a multi-index $\mathbf{j} \in \mathbb{N}_0^\ell$ with $2 \leq |\mathbf{j}| \leq k$, where $|\mathbf{j}|$ denotes the sum of the coordinates of $\mathbf{j}$. It should be chosen in such a way that the $\mathbf{j}$-th coefficient $\alpha_{\mathbf{j}}$ of P is nonzero.

Let $v \geq 1$ and u be any pair of relatively prime integers satisfying $|\alpha_{\mathbf{j}} v - u| \leq v^{-1}$, their existence being guaranteed by Dirichlet's Approximation Theorem.

Then we have the following Weyl sum estimate.

Theorem 2 ([15, Thm. 10.1])

$$\sum_{\mathbf{q} \leq Q} e(P(\mathbf{q})) \ll Q^{\ell+\varepsilon} \left(Q^{-1} + v^{-1} + Q^{-|\mathbf{j}|} v \right)^{\sigma'}.$$

Here the last bracket expression denotes the gain in comparison to the trivial estimate Q^ℓ. A near-optimal bound in the one-variable case has also been given by Wooley in [21] with the exponent $1/2k(k-1)$. It is clear that the classical approach by Montgomery in [14] now gives the bound with the exponent $\sigma = 1/k(k-1)$; one just needs to insert the main conjecture of VMVT due to [5, 22, 23] into the proof there. Thus, with almost no extra work, we have now

Theorem 3 ([14, § 4.1] with [5, 22, 23])

$$\sum_{q \leq Q} e(P(q)) \ll Q^{1+\varepsilon} \left(Q^{-1} + v^{-1} + Q^{-k} v \right)^{1/k(k-1)}.$$

In [10], we proved the following transformation of the bound from Theorem 2. The analogous theorem in the one-dimensional case can already be found in [9].

Theorem 4 ([10, Cor. 4.3])

$$\sum_{\mathbf{q} \leq Q} e(P(\mathbf{q})) \ll Q^{\ell+\varepsilon} \left(Q^{-1} + \sum_{1 \leq w \leq Q} \min \left(w^{-1}, Q^{-|\mathbf{j}|} \| w \alpha_{\mathbf{j}} \|^{-1} \right) \right)^{\sigma'},$$

where $\| \alpha \|$ denotes the distance of α to the nearest integer.

This bound was the one that was useful in a spacing problem with clusters of fractions with polynomial denominators, which is needed for the proof of polynomial LSIs. The original bound in Theorem 2 has not the right shape for this application.

The summand Q^{-1} that occurs in all bounds weakens the result. One might conjecture that the bound is also true without this summand. This observation was already made by Montgomery in [14, Conj. 1 in §3.2]. It is an open and difficult problem whether this summand can be eliminated.

Now we take a closer look at the aforementioned spacing problem with clusters of fractions. It is the following fraction counting problem: Give good upper bounds for the number of fractions of the shape $\frac{a}{P(\mathbf{q})}$ near a given fixed fraction $\frac{b}{P(\mathbf{r})} \neq 0$.

Assume that all considered $P(\mathbf{q})$ are positive. For $x > 0$ let

$$\mathscr{F}_{b, P(\mathbf{r})}(x) := \{(a, \mathbf{q}) \in \mathbb{Z} \times \mathbb{N}^{\ell}; Q < \mathbf{q} \leq 2Q, \left| \frac{a}{P(\mathbf{q})} - \frac{b}{P(\mathbf{r})} \right| \leq x\}.$$

Using a familiar Fourier analytic approach, commonly used for polynomial LSIs, we arrive at

Theorem 5 ([10, Eq. (8)])

$$\#\mathscr{F}_{b, P(\mathbf{r})}(x) \ll B^{-1}Q + B^{-1} \sum_{a \leq B} \left| \sum_{\mathbf{q} \leq 2Q} e\left(\frac{ab}{P(\mathbf{r})} P(\mathbf{q}) \right) \right|$$

with $B := (2M_Q x)^{-1}$ and $M_Q := \max\{P(\mathbf{q}); Q < \mathbf{q} \leq 2Q\}$.

As a normalization, assume that $\alpha_{\mathbf{j}} = 1$ is the coefficient in P satisfying $2 \leq |\mathbf{j}| \leq k$. Then the bound in Theorem 5 contains a Weyl sum with $\mathbf{j}$-th coefficient $\frac{ab}{P(\mathbf{r})}$.

The adjusted estimate for Weyl sums, Theorem 4, applies together with Hölder's inequality. This yields the following satisfying bound in the fraction counting problem.

Theorem 6 ([10, Thm. 2.2])

$$\#\mathscr{F}_{b, P(\mathbf{r})}(x) \ll B^{-1}Q + Q^{\ell - \sigma' + \varepsilon} + Q^{\ell + \varepsilon} P(\mathbf{r})^{-\sigma'} + B^{-\sigma'} Q^{\ell - |\mathbf{j}|\sigma' + \varepsilon} P(\mathbf{r})^{\sigma'}.$$

The one-dimensional versions of Theorems 4–6 already occur in [9], but with the exponent $1/2k(k - 1)$. By the main conjecture in VMVT, proved in [5, 22, 23], the one-dimensional cases of Theorems 4 and 6 can now be stated with the exponent $1/k(k - 1)$ instead of $1/2k(k - 1)$.

Then, the mentioned polynomial LSI bound (5) can be deduced from the above bound for $\#\mathscr{F}_{b, P(\mathbf{r})}(x)$ from Theorem 6.

4   Fraction Counting: Bounds for the Spacing Problem

In this section, we take a closer look at distances of fractions with k-th power denominator and discuss the impact that the polynomial LSI has on the estimate of fraction distances. We restrict to the one-dimensional case, where we have the optimal exponent $\sigma = 1/k(k-1)$ at hand.

Consider the set of pairs $\mathscr{F} := \{(a, q^k); \gcd(a, q) = 1, \ Q < q \leq 2Q\}$.

Let $(b, r^k) \in \mathscr{F}$ be fixed and for a pair $(a, q^k) \in \mathscr{F}$, consider the distance

$$D := D(a, q^k) := \frac{a}{q^k} - \frac{b}{r^k}.$$

Using the one-dimensional version of Theorem 6 and a few calculation steps, for any $N \geq 1$, we deduce the upper bound

$$\int_{1/N}^{1/2} \left(\sum_{\substack{(a,q^k)\in\mathscr{F} \\ |D|\leq x}} 1 \right) \frac{dx}{x^2} \ll \left(Q^{k+1} + N Q^{1-\sigma} + Q^{1+k\sigma} N^{1-\sigma} \right) (NQ)^\varepsilon. \qquad (7)$$

From this, we obtain estimates for the number of fractions with $|D|$ in specified intervals, as follows.

To orient the reader, we start by establishing a lower bound for $|D|$ using Roth's Theorem. It yields that for every real constant $\delta > 0$ and different fractions, we have the ineffective inequality

$$|D| = \left| \frac{a}{q^k} - \frac{b}{r^k} \right| = bq^{-k} \left| \frac{q}{r} - \sqrt[k]{\frac{a}{b}} \right|$$

$$\cdot \left| \left(\frac{q}{r} \right)^{k-1} + \left(\frac{q}{r} \right)^{k-2} \sqrt[k]{\frac{a}{b}} + \cdots + \left(\sqrt[k]{\frac{a}{b}} \right)^{k-1} \right|$$

$$\gg_{k,a,b} Q^{-k} Q^{-2-\delta} = Q^{-k-2-\delta},$$

supposing that $\sqrt[k]{\frac{a}{b}}$ is irrational. This result shows that most fractions are further away than $Q^{-k-2-\delta}$ from an arbitrary, single fraction, and gives some heuristic evidence that they are quite well-spaced.

As a consequence of the above upper bound (7), we will now show that we gain much more information on how these fractions are spaced. Namely, for λ close to 1, most fractions still have a distance bigger than $Q^{-k-\lambda}$ (a distance that goes beyond Roth's theorem), and only a few will be closer than that. An approach using the classical LSI might show this with the distance Q^{-k-1}, but the polynomial LSI gives information even beyond that.

In order to see this in detail, we split the inner sum in the integral in (7) according to if $|D| \geq 1/N$ holds or not. This yields

$$\int_{1/N}^{1/2} \left(\sum_{\substack{(a,q^k)\in\mathscr{F} \\ |D|\leq 1/N}} 1 \right) \frac{dx}{x^2} + \sum_{\substack{(a,q^k)\in\mathscr{F} \\ 1/N<|D|}} \int_{|D|}^{1/2} \frac{dx}{x^2} \gg AN + \frac{A_\lambda}{|D_\lambda|},$$

where

$$A := \#\{(a, q^k) \in \mathscr{F};\ |D| \leq 1/N\},$$

$$A_\lambda := \#\{(a, q^k) \in \mathscr{F};\ Q^{-k-\lambda} \geq |D| > 1/N\},$$

$$|D_\lambda| := \max\{|D|;\ Q^{-k-\lambda} \geq |D| > 1/N\},$$

supposing $N \geq Q^{k+\lambda}$ with $0 < \lambda \leq k$. We will keep this condition in the following consideration.

We aim to estimate A and A_λ, and due to the right-hand side in the bound (7), we have to distinguish the two cases $Q^{k+1} \leq N$ and $Q^{k+1} \geq N$.

Consider the first case, when $Q^{k+1} \leq N$. In this region, the upper bound in (7) is $\ll NQ^{1-\sigma}(NQ)^\varepsilon$, and we arrive at $AN \ll Q^{1-\sigma}N(NQ)^\varepsilon$, so $A \ll Q^{1-\sigma+\varepsilon}$. We conclude that the number of very close fractions is small: For at most $\ll Q^{1-\sigma+\varepsilon}$ many pairs $(a, q^k) \in \mathscr{F}$, we have $|D| \leq 1/N \leq Q^{-k-1}$.

Further, the number of close, but not very close fractions, is also small: If $1 \leq \lambda < k$, then $\frac{A_\lambda}{|D_\lambda|} \ll Q^{1-\sigma}N(NQ)^\varepsilon$. We obtain that the number of all pairs $(a, q^k) \in \mathscr{F}$ such that $1/N < |D| \leq Q^{-k-\lambda}$ is

$$A_\lambda \ll |D_\lambda|Q^{1-\sigma}N(NQ)^\varepsilon \ll Q^{-k-\lambda+1-\sigma}N(NQ)^\varepsilon \ll Q^{1+(1-\lambda)-\delta+\varepsilon} \ll Q^{1-\delta+\varepsilon},$$

supposing $N \ll Q^{k+1+\sigma-\delta}$, so that $\lambda \leq 1 + \sigma - \delta$.

Concerning far fractions, we know that there are $\gg Q^2$ pairs with bigger $|D|$.

Now consider the second case, when $N \leq Q^{k+1}$, so that $0 < \lambda \leq 1$.

In this region, the upper bound in (7) is $\ll Q^{1+k\sigma}N^{1-\sigma}(NQ)^\varepsilon$, which yields $A \ll Q(Q^k/N)^{\sigma+\varepsilon}$. So the number of very close fractions is small: For at most $\ll Q(Q^k/N)^{\sigma+\varepsilon} \ll Q^{1-\lambda\sigma+\varepsilon}$ many pairs $(a, q^k) \in \mathscr{F}$, we have $|D| \leq 1/N$.

Further, the number of close, but not very close fractions, is also small: If $0 < \lambda \leq 1$, then $\frac{A_\lambda}{|D_\lambda|} \ll Q^{1+k\sigma}N^{1-\sigma}(NQ)^\varepsilon$, so the number of all pairs $(a, q^k) \in \mathscr{F}$ such that $1/N < |D| \leq Q^{-k-\lambda}$ is

$$A_\lambda \leq |D_\lambda|Q^{1+k\sigma}N^{1-\sigma}(NQ)^\varepsilon \ll Q^{-k-\lambda}(N/Q^k)^{1-\sigma}Q^{k+1}(NQ)^\varepsilon \ll Q^{1-\delta(1-\sigma)+\varepsilon},$$

supposing $N \ll Q^{k+\lambda/(1-\sigma)-\delta}$.

Concerning far fractions, we know as in the first case that there are $\gg Q^2$ pairs with bigger $|D|$.

All these conclusions depend directly on the value σ, the exponent in VMVT.

5   Consequences of the Polynomial LSI for the Distribution of Primes in Certain Arithmetic Progressions

A well-known important application of the standard LSI is that it leads to a proof of the celebrated theorem of E. Bombieri and A. I. Vinogradov from 1965/1966. The modern proof in most textbooks that uses the standard LSI goes back to R. C. Vaughan. A recommendable presentation can be found in [18].

Theorem 7 ([6, 19, 20]) *For real $A, Q, x > 1$ we have*

$$\sum_{q \leq Q} E(x, q) \ll_A \frac{x}{(\log x)^A} \tag{8}$$

with

$$E(x, q) := \max_{\substack{a \bmod q \\ \gcd(a,q)=1}} \left| \sum_{\substack{n \leq x \\ n \equiv a \bmod q}} \Lambda(n) - \frac{x}{\varphi(q)} \right|, \tag{9}$$

provided that $Q \leq x^{1/2}(\log x)^{-3-A}$.

This theorem is far-reaching in number theory since it can replace the assumption of Riemann's hypothesis in many applications.

A deep conjecture concerning Bombieri–Vinogradov's theorem is Elliott–Halberstam's conjecture in [8]. It states that $1/2$, the exponent in the bound for Q in the theorem, can be replaced by $1 - \varepsilon$. In fact, proving the theorem for any exponent bigger than $1/2$ would already be a big improvement.

The celebrated breakthrough of Zhang [24] on bounded gaps between primes relies on a numerical improvement in this exponent that is possible for certain moduli being smooth numbers that suffice for this application. Therefore it is of interest to study variants with moduli coming from suitable sets of integers.

Several such variants of Bombieri–Vinogradov's theorem have been proved and applied in the past. Already for the obvious variant of moduli in polynomial shape, there exists a vast literature. In [4], R.C. Baker recently gave a proof that for $P(x) = x^2$, we have

$$\sum_{Q < q \leq 2Q} \frac{\varphi(q^2)}{q} E(x, q^2) \ll_A \frac{x}{(\log x)^A}$$

for $Q \ll x^{1/4-\varepsilon}$, which can be considered as the best one can obtain using current methods. He uses the bound (4) by Baier and Zhao [2]. Here the factor $\frac{\varphi(q^2)}{q}$ respects the sparsity of the set of squares; we included it here to follow the notation by Elliott in [7]. In the general case, the current region for Q is $Q \ll x^{9/20d-\varepsilon}$ by Baker in [3], but we expect that $Q \ll x^{1/2d-\varepsilon}$ should be the correct region for Q.

The present literature usually focuses on moduli of polynomial shape in one variable, which is already difficult enough. Moduli that are smooth numbers as in Zhang's version might be seen as values of monomials $x_1 \cdots x_r$, where the variables x_i run through integers up to a suitable bound depending on Q.

One might expect that further multidimensional variants are of interest. A very special multivariable version, that shows that we can obtain more control over the moduli in certain cases, has just recently been published by the author in [11], but appears to be too complicated and still too weak for applications. We present some of the main ideas that have led to this version.

The central problem to deal with is to find suitable multivariable polynomials for which Vaughan's approach applies at all using the polynomial LSI instead of the standard LSI. Trying to adjust Vaughan's proof of Bombieri–Vinogradov's theorem, it turns out that to make it work, the polynomials should quite often attain squarefree values.

The idea is to take the values of a polynomial of the following special shape as moduli: Let $P \in \mathbb{Z}[x_1, \ldots, x_\ell]$ denote the polynomial

$$P(\mathbf{x}) = \left(x_{u(1)}^2 + x_{v(1)}^2\right) \cdot \left(x_{u(2)}^2 + x_{v(2)}^2\right) \cdots \left(x_{u(k-1)}^2 + x_{v(k-1)}^2\right) \cdot \left(x_{u(k)}^2 + x_{v(k)}^2\right),$$
(10)

where u, v are suitable maps from $\{1, \ldots, k\}$ to $\{1, \ldots, \ell\}$. This is a polynomial in ℓ variables of degree $2k$, and since sums of two squares often are primes (namely, the primes $\equiv 1 \bmod 4$; they are norms of prime elements of $\mathbb{Z}[i]$, so they can uniquely be written as a sum of two squares), these polynomials attain a lot of squarefree values. Further, the construction with u, v allows them to have a large degree $2k$ compared to a rather small number ℓ of variables, so that we can expect nontrivial bounds from the polynomial LSI. But such a suitable polynomial P in (10) is of a strange shape, since variables have to be repeated. Pairs or tuples of primes $\equiv 1$ mod 4 that share a square in their representation as a sum of two squares are not much explored.

In this way, we obtain the following theorem in the style of Bombieri–Vinogradov with products of primes $\equiv 1$ mod 4 as moduli, where the estimate respects the sparsity of the moduli set:

Theorem 8 ([11, Thm. 1.2]) *If* $x^{\varepsilon/\sigma'} \ll Q \le x^{\frac{1/3-2\varepsilon}{2k-\sigma'}}$ *we have*

$$\sum_{Q < \mathbf{q} \le 2Q} G_{\mathbf{q}} \frac{\varphi(P(\mathbf{q}))}{Q^\ell} E(x, P(\mathbf{q})) \ll \frac{x}{\log^A(x)},$$

where

$$G_{\mathbf{q}} = \mu^2(P(\mathbf{q})) \Lambda \left(q_{u(1)}^2 + q_{v(1)}^2 \right) \cdots \Lambda \left(q_{u(k)}^2 + q_{v(k)}^2 \right).$$

Here, $\sigma' = \sigma'(2k, \ell)$ is the exponent from VMVT in (6). So the exponent of x for the region of Q is mainly $1/6k$ improved by a term depending on σ'. So to speak, the result breaks a $1/6k$-"barrier" by an amount depending on the exponent σ' due to the work around VMVT, and this is done with little effort. But it is unclear whether this is the right barrier. One might rather expect that $1/4k$ is the correct exponent of validity feasible using current methods (since $4k$ is the double of the degree of P).

A key step in the proof of this theorem is a multivariable polynomial version of the so-called basic mean value theorem due to Vaughan, see [18]. This theorem is an intermediate result in the proof of Bombieri–Vinogradov's theorem and is also commonly known as Vaughan's inequality. It states the following.

Theorem 9 ([18, Basic Mean Value Theorem]) *For $M \geq 1$ and $x \geq 2$ we have*

$$\sum_{q \leq Q} \frac{q}{\varphi(q)} \sum_{\chi(q)}^{*} \sup_{y \leq x} \left| \sum_{n \leq y} \Lambda(n) \chi(n) \right| \ll \left(x + x^{5/6} Q + x^{1/2} Q^2 \right) \log^3(xQ).$$

As a generalization, in [11] we established the following polynomial basic mean value theorem, which is valid for arbitrary suitable polynomials P. Let σ' be the corresponding exponent from VMVT. Then we have the following theorem.

Theorem 10 ([11, Thm. 4.1])

$$\sum_{Q < \mathbf{q} \leq 2Q} \frac{P(\mathbf{q})}{\varphi(P(\mathbf{q}))} \sum_{\chi(P(\mathbf{q}))}^{*} \sup_{y \leq x} \left| \sum_{n \leq y} \Lambda(n) \chi(n) \right| \ll \Delta_0(Q, x)(xQ)^{\varepsilon}$$

with

$$\Delta_0(Q, x) := Q^{\ell - \sigma'} x + Q^{\ell + (k - \sigma')/2} x^{5/6} + Q^{\ell + (k-1)\sigma'/2} x^{1 - \sigma'/6},$$

provided that $x \geq Q^{2k + \sigma'}$.

Very likely, Theorem 10 does not yet give the best possible bound; here, further methods are needed to improve the ranges. A direct path to do this would be to give improvements of the bound in the polynomial LSI.

Zhang's proof on bounded gaps between primes in [24] has much been simplified by the polymath projects [17]. A more accessible proof with numerical advantage was published by Maynard in [12], and moreover, it applies to prime k-tuples in general. Likewise, Maynard's method shows that the problem to establish a good numerical bound for C such that infinitely many prime pairs of distance at most C exist relies heavily on the exponent in Bombieri–Vinogradov's theorem.

A suggestion to a path for further numerical improvements towards the twin prime conjecture could be the use of variants of Bombieri–Vinogradov's theorem with polynomial moduli, since with these, much deeper estimates can be at hand. In Maynard's proof, squarefree moduli play a crucial role due to the use of multidimensional Selberg sieve weights, so the focus should be on polynomials with many squarefree values. The suggested polynomials in (10) may do the job with adjusted Selberg sieve weights, but are difficult to handle due to their strange shape, and probably Theorem 8 in the current state is not sharp enough to hope for improvements.

Acknowledgements The author thanks the referee for many valuable comments and suggestions. Thanks are due to the Fields Institute in Toronto for the support to take part at the workshop on Efficient Congruencing and Translation-invariant Systems in March 2017.

References

1. S. Baier, L. Zhao, Large sieve inequality with characters for powerful moduli. Int. J. Number Theory **1**(2), 265–279 (2005)
2. S. Baier, L. Zhao, An improvement for the large sieve for square moduli. J. Number Theory **128**(1), 154–174 (2008)
3. R.C. Baker, Primes in arithmetic progressions to spaced moduli. Acta Arith. **153**(2), 133–159 (2012)
4. R.C. Baker, Primes in arithmetic progressions to spaced moduli. III. Acta Arith. **179**(2), 125–132 (2017)
5. J. Bourgain, C. Demeter, L. Guth, Proof of the main conjecture in Vinogradov's mean value theorem for degrees higher than three. Ann. Math. (2) **184**(2), 633–682 (2016)
6. E. Bombieri, On the large sieve. Mathematika **12**, 201–225 (1965)
7. P.D.T.A. Elliott, Primes in short arithmetic progressions with rapidly increasing differences. Trans. Am. Math. Soc. **353**(7), 2705–2724 (2001)
8. P.D.T.A. Elliott, H. Halberstam, A conjecture in prime number theory, in *Symposia Mathematica (INDAM, Rome, 1968/69)*, vol. IV (Academic, London, 1970), pp. 59–72
9. K. Halupczok, A new bound for the large sieve inequality with power moduli. Int. J. Number Theory **8**(3), 689–695 (2012)
10. K. Halupczok, Large sieve inequalities with general polynomial moduli. Q. J. Math. **66**(2), 529–545 (2015)
11. K. Halupczok, A Bombieri–Vinogradov theorem with products of Gaussian primes as moduli. Funct. Approx. Comment. Math. **57**, 77–91 (2017). Advance Publication
12. J. Maynard, Small gaps between primes. Ann. Math. (2) **181**(1), 383–413 (2015)
13. H.L. Montgomery, The analytic principle of the large sieve. Bull. Am. Math. Soc. **84**(4), 547–567 (1978)
14. H.L. Montgomery, *Ten Lectures on the Interface Between Analytic Number Theory and Harmonic Analysis*, CBMS Number 84 (1994)
15. S.T. Parsell, S.M. Prendiville, T.D. Wooley, Near-optimal mean value estimates for multidimensional Weyl sums. Geom. Funct. Anal. **23**(6), 1962–2024 (2013)
16. L.B. Pierce, The Vinogradov mean value theorem [after Wooley, and Bourgain, Demeter and Guth]. arXiv:1707.00119 [math.NT]
17. Polymath-projects 8a and 8b: Bounded gaps between primes, http://michaelnielsen.org/polymath1/index.php?title=Bounded_gaps_between_primes

18. R.C. Vaughan, *The Bombieri–Vinogradov Theorem*, AIM discussion paper, November 2005. Available at http://www.personal.psu.edu/rcv4/Bombieri.pdf
19. A.I. Vinogradov, On the density hypothesis for Dirichlet L–series. Izv. Akad. Nauk SSSR, Ser. Mat. **29**, 903–934 (1965)
20. A.I. Vinogradov, Corrections to the work of A. I. Vinogradov 'On the density hypothesis for Dirichlet L–series'. Izv. Akad. Nauk SSSR, Ser. Mat. **30**, 719–729 (1966)
21. T. Wooley, Vinogradov's mean value theorem via efficient congruencing. Ann. Math. (2), **175**(3), 1575–1627 (2012)
22. T. Wooley, The cubic case of the main conjecture in Vinogradov's mean value theorem. Adv. Math. **294**, 532–561 (2016)
23. T. Wooley, Nested efficient congruencing and relatives of Vinogradov's mean value theorem. arXiv:1708.01220v2 [math.NT]
24. Y. Zhang, Bounded gaps between primes. Ann. Math. (2) **179**(3), 1121–1174 (2014)
25. L. Zhao, Large sieve inequality with characters to square moduli. Acta Arith. **112**(3), 297–308 (2004)

Unexpected Regularities in the Behavior of Some Number-Theoretic Power Series

A. J. Hildebrand

Abstract The goal of this paper is to draw attention to a surprising and little-known phenomenon, namely the unexpected *regularity* in the behavior of the Möbius power series $\sum_{n=1}^{\infty} \mu(n)z^n$, and some related series. This phenomenon was first pointed out and investigated a half century ago in a remarkable, but now nearly forgotten, paper by Carl-Erik Fröberg. Its manifestations include "fake" asymptotics as $z \to 1$, and error terms that are significantly better than the usual error terms in prime number estimates. We describe these results and some recent developments, explain the underlying phenomenon, and comment on possible applications.

1 Introduction

Let $\mu(n)$ denote the Möbius function and consider the associated power series

$$f(z) = \sum_{n=1}^{\infty} \mu(n)z^n. \tag{1}$$

Clearly, $f(z)$ converges when $|z| < 1$, and diverges when $|z| \geq 1$. It is therefore of interest to consider the behavior of $f(z)$ as z approaches the unit circle. In particular, convergence of $f(r)$ as $r \to 1-$ through real values would mean that the series $\sum_{n=1}^{\infty} \mu(n)$ is Abel summable.

Anyone with some experience in estimating sums involving prime numbers or the Möbius function would likely be quite skeptical about such a conclusion: That $\sum_{n=1}^{\infty} \mu(n)$ is Abel summable seems too good a result to be true, given the known irregularities in the distribution of primes and the squareroot type oscillations in the summatory function $\sum_{n \leq x} \mu(n)$. Indeed a back-of-the-envelope calculation using partial summation suggests that these oscillations translate into

A. J. Hildebrand (✉)
Department of Mathematics, University of Illinois, Urbana, IL, USA
e-mail: ajh@illinois.edu

© Springer International Publishing AG, part of Springer Nature 2018   111
J. Pintz, M. Th. Rassias (eds.), *Irregularities in the Distribution of Prime Numbers*,
https://doi.org/10.1007/978-3-319-92777-0_6

Table 1 Numerical values of the Möbius power series $f(r) = \sum_{n=1}^{\infty} \mu(n)r^n$ as $r \to 1-$

r	$f(r)$	Source
0.9	−1.1384289	[5]
0.99	−1.8867855	[5]
0.999	−1.9881049	[5]
0.9999	−1.9988015	[5]
0.99999	−1.9998804	[5]
0.999999	−1.9999878	[3]
0.9999999	−1.9999945	[3]

corresponding unbounded oscillations of $f(r)$ unless some miraculous cancellation between positive and negative terms were to occur.

Yet, if one computes the series $f(r)$ numerically for values r close to 1, one is in for a surprise: Rather than observing a random behavior with large oscillations, the numerical data, shown in Table 1, seem to provide compelling evidence that $f(r)$ does indeed converge as $r \to 1-$, with limit -2. In other words, the empirical data points towards $\sum_{n=1}^{\infty} \mu(n)$ being in fact Abel summable, with -2 as the Abel sum.

Further support for this conclusion is provided by the Dirichlet series

$$\sum_{n=1}^{\infty} \frac{\mu(n)}{n^s} = \frac{1}{\zeta(s)}.$$

Formally substituting $s = 0$ into this series leads to the heuristic

$$\sum_{n=1}^{\infty} \mu(n) \doteq \frac{1}{\zeta(0)} = \frac{1}{-1/2} = -2,$$

with the same value -2 now arising as the "Dirichlet sum" of $\sum_{n=1}^{\infty} \mu(n)$.

Alas, all this turns out to be a mirage as it contradicts a known oscillation result:

Theorem 1.1 (Delange [4], Katai [8]) *As* $r \to 1-$, *we have*

$$f(r) = \Omega_{\pm}\left(\frac{1}{\sqrt{1-r}}\right). \tag{2}$$

In particular, (2) implies that $f(r)$ is unbounded in both positive and negative direction, so $\sum_{n=1}^{\infty} \mu(n)$ is very far from being Abel summable.

Why is there such an apparent contradiction between the numerical data and the known theoretical results? The answer can be found in a remarkable, but now nearly forgotten, 1966 paper by Fröberg [5]. Fröberg provided extensive numerical data on the behavior of $f(z)$ and also gave a heuristic formula for $f(r)$ explaining both the data in Table 1 and the asymptotic behavior indicated in (2).

The rest of this paper is organized as follows. In Sect. 2 we present the Fröberg heuristic on the behavior of $f(r)$. In Sect. 3 we describe another manifestation of the "unreasonable" regularity in the behavior of the Möbius power series, namely better than expected upper bounds for $f(r)$; this is based on a 2010 paper by Gerhold [6]. In Sect. 4 we report on recent numerical investigations [3] on Fröberg-like behavior in more general number-theoretic power series. We conclude in Sect. 5 with some remarks on related questions and possible applications.

2   Fake Asymptotics and the Fröberg Heuristic

In 1966 Carl-Erik Fröberg published a paper [5] titled "Numerical Studies of the Möbius Power Series," in the journal *Nordisk Tidskr. Informations-Behandling (BIT)*. The paper contains a detailed numerical investigation of the series $f(z) = \sum_{n=1}^{\infty} \mu(n)z^n$ and $g(z) = \sum_{n=1}^{\infty} \mu(n)z^n/n$, along with heuristics explaining some of the observed behavior. It represents an impressive tour de force in computational number theory for its time, both for the amount and quality of numerical data it contains, and for the exceptional amount of care and detail the author has devoted to deriving a heuristic for $f(z)$. The main heuristic, stated below, gives an explicit expansion of $f(e^{-t})$ involving nine leading terms, with numerical coefficients that have been worked out to six or more digit accuracy.

Fröberg's paper seems to have received little attention within the mathematical community, perhaps because it was published in a relatively obscure journal. MathSciNet lists no citations to this paper, while Google Scholar lists two citations in journal articles, one a 1995 paper by Bohman and Fröberg [2], and the other a paper by Bateman and Diamond [1] who proved an oscillation result similar to (2) from Tauberian theorems, and commented on the "fake" asymptotic phenomenon observed by Fröberg.

Fröberg starts out his analysis by representing $f(e^{-t})$ (where $t > 0$) in terms of Mellin transforms:

$$f(e^{-t}) = \frac{1}{2\pi i} \int_{c-i\infty}^{c+i\infty} \frac{\Gamma(s)}{\zeta(s)} t^{-s}\, ds, \tag{3}$$

where $c > 1$. Shifting the path of integration to the left then leads to an expression for $f(e^{-t})$ as a sum over residue contributions from the poles of the integrand.

Fröberg notes that there are three sources for such poles: (1) poles of $\Gamma(s)$ at $s = 0, -1, -3, -5, \ldots$ at which $\zeta(s)$ is nonzero; (2) poles of $\Gamma(s)$ at $s = -2, -4, \ldots$, which are also zeros of $\zeta(s)$; and (3) non-trivial zeros of $\zeta(s)$. He then works out in detail the residues corresponding to each of these three types of poles. The pole at $s = 0$ contributes a residue -2, while the poles at $s = -(2m + 1)$, $m = 1, 2, \ldots$, contribute residues that are of the form ct^{2m+1}. The poles at $s = -2m$, $m = 1, 2, \ldots$, contribute terms that are linear combinations of t^{2m} and $t^{2m} \log t$. Finally, the poles due to non-trivial zeta zeros contribute residues of the form $ct^{-1/2} \cos(a + b \log(1/t))$, with appropriate constants a, b, c. Fröberg calculates the constants involved in high accuracy.

The upshot of these calculations, as presented by Fröberg in his paper [5], is the following heuristic expansion:

Heuristic 2.1 (Fröberg [5, (4.6)]) *As $t \to 0+$, we have*

$$f(e^{-t}) = -2 + 12t + 2.58027981956t^2 - 20t^3 + 5.98503868t^4 \qquad (4)$$

$$+ \cdots + 16.42119337t^2 \log t - 5.2188941247t^4 \log t$$

$$+ \cdots + \frac{1.439517 \times 10^{-9}}{\sqrt{t}} \cos\left(4.298514 + 14.134725 \log(1/t)\right) + \cdots$$

The last term in (4) represents the contribution of the first pair of non-trivial zeros of $\zeta(s)$ (which occurs at a height of approximately 14.13) and the ellipses at the end represent terms corresponding to higher-lying zeta zeros.

Of the nine terms on the right side of this expansion, all tend to 0 as $t \to 0$, except for the constant term, -2, and the oscillating terms at the end. Thus a simplified version of (4) can be stated as

$$f(e^{-t}) \approx -2 + \frac{1.439517 \times 10^{-9}}{\sqrt{t}} \cos\left(c_1 + c_2 \log(1/t)\right), \qquad (5)$$

where c_1, c_2 are numerical constants.

The crucial feature of (5) is the extremely small coefficient attached to the term involving $1/\sqrt{t}$, which causes this term to be negligible compared to the constant term -2 unless t is of order 10^{-16} or smaller. In particular, in the range covered by the numerical data in Table 1, which corresponds to values $t \gtrsim 10^{-7}$, this term is of order 10^{-6} and hence essentially undetectable in the numerical data. By contrast, for t-values of around 10^{-18} or smaller the last term in (5) begins to take over and dominate the behavior of $f(r)$.

This explains the apparent discrepancy between numerical data (suggesting convergence to -2 as $t \to 0$) and theoretical results (showing that $f(e^{-t})$ is unbounded, with oscillations of order $1/\sqrt{t}$). Figure 1 illustrates this clearly, showing both the numerical data for the range $t > 10^{-6}$ and the heuristic (5).

The smallness of the coefficient of the last term in (5)—and hence the source of the "fake" asymptotic behavior of $f(r)$—is due to the exponential decay of the Gamma function along a vertical line, combined with the fact that the lowest non-trivial zero of $\zeta(s)$ occurs at the relatively large height of around 14.13. Indeed, the size of this coefficient is largely determined by the size of $|\Gamma(0.5 + i\beta)|$, where β is the height of the associated zeta zero. With $\beta \approx 14.13$, we have $|\Gamma(0.5 + i\beta)| \approx 5.7 \times 10^{-10}$.

Mellin transform representations analogous to (3) for other weighted sums of the Möbius function such as $\sum_{n \le x} \mu(n)$, $\sum_{n \le x} \mu(n)(x - n)$, or $\sum_{n \le x} \mu(n) \log(x/n)$, all involve integrands with a polynomially decaying function in place of the Gamma function. Thus, oscillations due to zeta zeros maintain nearly their full effect on the behavior of these sums. As a result, these sums do not display the "fake" asymptotic behavior observed in the Möbius power series.

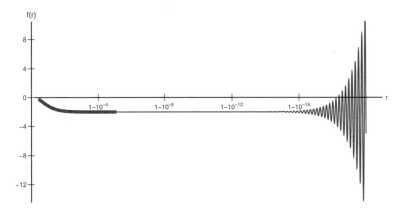

Fig. 1 Behavior of the Möbius power series $f(r) = \sum_{n=1}^{\infty} \mu(n)n^r$ as $r \to 1$: The thick solid line shows the *actual* values of $f(r)$ within the computable range, while the thin line shows the *predicted* long-range behavior given by the right-hand side of (5)

3 Better than Expected Error Terms

The rapid decay of the Gamma function in the integral (3) that is the root cause of the "fake" asymptotic behavior of the Möbius power series studied by Fröberg also leads to better-than-expected error terms when estimating this series. This was observed by Gerhold [6], and what follows is based on his paper [6].

The best known unconditional bound for the partial sum of the Möbius function is

$$\sum_{n \leq x} \mu(n) \ll x \exp\left\{ -\frac{c(\log x)^{3/5}}{(\log \log x)^{1/5}} \right\}, \tag{6}$$

representing a saving of $O(\exp\{-(\log x)^{3/5+o(1)}\})$ over the trivial bound $O(x)$. This is the expected type of saving in estimates of sums involving prime numbers or the Möbius function, given our current knowledge on zero-free regions for the Riemann zeta function.

Using (6) and partial summation yields the following estimate for the Möbius power series $f(z) = \sum_{n=1}^{\infty} \mu(n)z^n$:

$$f(e^{-t}) \ll \frac{1}{t} \exp\left\{ -\frac{c(\log(1/t))^{3/5}}{(\log \log(1/t))^{1/5}} \right\}. \tag{7}$$

The saving over the trivial bound $O(1/t)$ here is of the same type as in (6), with the familiar exponent $3/5$ in the logarithm in the exponent.

Starting from the integral representation (3) and exploiting the rapid decay of the Gamma function in this representation, Gerhold was able to significantly improve over the estimate (7), essentially replacing the exponent $3/5$ of $\log(1/t)$ by 1:

Theorem 3.1 (Gerhold [6, Theorem 2]) *As $t \to 0+$, we have*

$$f(e^{-t}) \ll \frac{1}{t} \exp\left\{-\frac{0.0203 \log(1/t)}{(\log\log(1/t))(\log\log\log(1/t))^{1/3}}\right\}. \tag{8}$$

In fact, Gerhold proved a result of this type for more general power series $\sum_{n=1}^{\infty} a_n z^n$, which yields estimates with similar better-than-expected error terms for a large class of power series with number-theoretic coefficients. For example (see [6, Corollary 5]), the power series with coefficients given by the Liouville function, $\lambda(n)$, and the von Mangoldt function, $\Lambda(n)$, satisfy the estimates

$$\sum_{n=1}^{\infty} \lambda(n)z^n = O(E(z)),$$

$$\sum_{n=1}^{\infty} \Lambda(n)z^n = \frac{1}{1-z} + O(E(z)),$$

as $z \to 1$ in an arbitrary sector of the form $|\arg(1-z)| \leq \pi/2 - \epsilon$, where $E(z)$ is the term on the right-hand size of (8) with $t = -\log z$.

4 Fake Asymptotics for Other Number-Theoretic Power Series

Do power series with "Möbius like" coefficients exhibit "fake" asymptotic behavior similar to that of the Möbius power series? Experience with analytic number theory estimates would suggest that minor changes such as replacing the Möbius function by the Liouville function (the completely multiplicative "cousin" of the Möbius function) or adding a coprimality condition (which amounts to changing the definition of the Möbius function at finitely many primes) should not materially affect the asymptotic behavior. This is indeed the case with the results of Gerhold described in the last section: The same "better than expected" estimate holds for the power series with the Liouville function and other "Möbius like" coefficients.

Surprisingly, when it comes to "fake" asymptotics, things are much more subtle, as shown in the recent work with Chen et al. [3]. In this section, we provide an overview of these findings.

Table 2 gives numerical values for four power series with "Möbius like" coefficients. As the table shows, "fake" asymptotic behavior appears to occur in seemingly unpredictable ways. On the one hand, some perturbations of the Möbius

Table 2 Numerical values of power series with coefficients $\mu(n)n$, $\mu(n)\chi_0(n)$, $\mu(n)\chi(n)$, and $\lambda(n)$ (where χ_0 and χ are the principal, respectively non-principal, Dirichlet characters modulo 3)

r	$\sum\limits_{n=1}^{\infty}\mu(n)nr^n$	$\sum\limits_{n=1}^{\infty}\mu(n)\chi_0(n)r^n$	$\sum\limits_{n=1}^{\infty}\mu(n)\chi(n)r^n$	$\sum\limits_{n=1}^{\infty}\lambda(n)r^n$
0.9	-5.898949	-1.255846	2.151627	-0.8725
0.99	-10.692559	-5.037126	2.847939	-5.05338
0.999	-11.795198	-9.183854	2.977714	-18.1857
0.9999	-11.984071	-13.370934	2.997442	-59.6842
0.99999	-11.693028	-17.562186	3.002186	-190.905
Apparent limit	-12	???	3	???

function, such as multiplying $\mu(n)$ by n or by a non-principal Dirchlet character, do not seem to affect the "fake" asymptotic behavior except for the value of the (apparent) limit. On the other hand, other—seemingly equally inconsequential— changes, such as restricting the function to integers that are relatively prime to 3, cause the behavior of the power series to change completely.

In what follows we explain the behavior observed for each of the four types of power series in Table 2, and we provide a heuristic for each case that matches up well with the numerical data.

4.1 Möbius Series with "Scaled" Coefficients

The "fake" convergence in the first column in Table 2, involving the "scaled" Möbius function $\mu(n)n$, is easiest to explain: Replacing $\mu(n)$ by $\mu(n)n$ amounts to replacing the generating Dirichlet series,

$$\sum_{n=1}^{\infty} \frac{\mu(n)}{n^s} = \frac{1}{\zeta(s)},$$

by

$$\sum_{n=1}^{\infty} \frac{\mu(n)n}{n^s} = \frac{1}{\zeta(s-1)}.$$

This changes the residue of the integrand in (3) at $s = 0$ from $1/\zeta(0) = -2$, the "fake" limit in the Möbius power series, to $1/\zeta(-1) = 1/(-1/12) = -12$. The latter quantity is indeed the apparent limit of the series $\sum_{n=1}^{\infty} \mu(n)nr^n$ in the numerical data of Table 2.

The same reasoning gives the following more general heuristic. Here, and in the remainder of this section, we use the symbols " $\rightarrow''$ " and " $\sim''$ " (with quotes) to denote "fake" convergence and "fake" asymptotic behavior, that is, asymptotic behavior that can be observed in the numerical data within the computable range, though does not necessarily reflect the true (i.e., long-term) behavior.

Heuristic 4.1 (Möbius Series with Scaled Coefficients) *Let* $-1 < \alpha \leq 1$. *As* $r \to 1-$, *we have within the computable range*

$$\sum_{n=1}^{\infty} \mu(n) n^{\alpha} r^n \text{ `` } \to \text{ '' } \frac{1}{\zeta(-\alpha)}.$$

4.2 Möbius Series with Non-principal Characters

A similar argument explains the "fake" convergence in the third column in Table 2. Replacing $\mu(n)$ by $\mu(n)\chi(n)$, where χ is a non-principal Dirichlet character, corresponds to replacing $1/\zeta(s)$ by $1/L(\chi, s)$, where $L(\chi, s)$ is the Dirichlet L-function associated with the character χ. The residue contribution at $s = 0$ then becomes $1/L(\chi, 0)$, provided $L(\chi, 0) \neq 0$. Thus we obtain the following heuristic.

Heuristic 4.2 (Möbius Series with Character Twist) *Let* χ *be a non-principal Dirichlet character such that the associated L-function is non-zero at* $s = 0$. *As* $r \to 1-$, *we have within the computable range*

$$\sum_{n=1}^{\infty} \mu(n) \chi(n) r^n \text{ `` } \to \text{ '' } \frac{1}{L(\chi, 0)}.$$

In the data of Table 2, χ is the non-principal Dirichlet character modulo 3. For this character we have $L(\chi, 0) = 1/3$, so the above heuristic suggests "fake" convergence behavior with limit $1/(1/3) = 3$. This is in excellent agreement with the numerical data in Table 2.

4.3 Möbius Series with Coprimality Condition

Multiplying the coefficient $\mu(n)$ by a *principal* Dirichlet character is equivalent to restricting n by a coprimality condition of the form $(n, q) = 1$. Surprisingly, this seemingly minor perturbation dramatically changes the (apparent) asymptotic behavior of the power series, as we will explain.

Consider the case shown in Table 2, i.e., a series with coefficients $\mu(n)\chi_0(n)$, where χ_0 is the principal Dirichlet character modulo 3, that is, the characteristic function of integers n satisfying $(n, 3) = 1$. On the Dirichlet series side, this amounts to replacing $\sum_{n=1}^{\infty} \mu(n) n^{-s} = 1/\zeta(s)$ by the Dirichlet series

Table 3 Numerical values of the power series with coefficients $\mu(n)\chi_0(n)$ (where χ_0 is the principal Dirichlet characters modulo 3) along with the heuristic given by (10)

r	$\displaystyle\sum_{n=1}^{\infty}\mu(n)\chi_0(n)r^n$	$3.39662 + 1.82048 \log\log(1/r)$
0.9	-1.25584617	-0.70012095
0.99	-5.03712686	-4.97784826
0.999	-9.18385465	-9.17788474
0.9999	-13.37093419	-13.37051089
0.99999	-17.56218604	-17.56239936

$$\sum_{n=1}^{\infty} \frac{\mu(n)\chi_0(n)}{n^s} = \prod_{p\neq 3}\left(1 - \frac{1}{p^s}\right) = \frac{1}{\zeta(s)(1 - 3^{-s})}. \tag{9}$$

With this change, the integrand in (3) becomes

$$\frac{\Gamma(s)t^{-s}}{\zeta(s)(1 - 3^{-s})},$$

with an extra factor $1 - 3^{-s}$ in the denominator. Since this factor has a zero at $s = 0$, the pole of the integrand at this point becomes a double pole, which completely changes the residue contribution at this point. Indeed, a routine but tedious computation gives

$$\mathrm{Res}\left(\frac{\Gamma(s)t^{-s}}{\zeta(s)(1 - 3^{-s})}, s = 0\right) = \frac{1}{\log 3}\left(2\gamma - \log 3 + 2\log(2\pi) + 2\log t\right) \tag{10}$$

$$\approx 3.39662 + 1.82048\log t,$$

where γ is Euler's constant. The crucial fact here is that the residue now is an unbounded function of t (and hence of $r = e^{-t}$) instead of being a constant. This explains why, in contrast to the above two cases, we do not see a "fake" convergence in this case.

Table 3 shows that the expression on the right of formula (10) represents a remarkably good approximation for the actual values of the series within the computable range.

The above reasoning can, in principle, be carried out for any Möbius series restricted by a coprimality condition of the form $(n, q) = 1$, yielding a heuristic for the behavior of this series within the computable range. However, the residue calculation gets quite complicated as the number of prime factors of q increases, so we state the result only for the case of prime moduli q:

Heuristic 4.3 (Möbius Series with Coprimality Condition) *Let p be an odd prime number. As $r \to 1-$, we have within the computable range*

$$\sum_{n=1,(n,p)=1}^{\infty} \mu(n)r^n \text{ `` } \sim \text{ " } -\frac{2}{\log p}\log\log(1/r)$$

4.4 The Liouville Power Series

We conclude this section by providing a heuristic explanation for the behavior observed in the final column in Table 2. The Dirichlet series associated with the Liouville function $\lambda(n)$ is

$$\sum_{n=1}^{\infty}\frac{\lambda(n)}{n^s} = \frac{\zeta(2s)}{\zeta(s)}.$$

Thus, the behavior of the Liouville power series, $\sum_{n=1}^{\infty}\lambda(n)r^n$, is determined by the residues of the function

$$\frac{\Gamma(s)\zeta(2s)t^{-s}}{\zeta(s)},$$

where $t = \log(1/r)$. The extra factor $\zeta(2s)$ does not cause any problems at the pole $s = 0$; the residue at this point is

$$\text{Res}\left(\frac{\Gamma(s)\zeta(2s)t^{-s}}{\zeta(s)}, s = 0\right) = 1. \tag{11}$$

On the other hand, the factor $\zeta(2s)$ causes the above function to have an additional pole at $s = 1/2$, with residue

$$\text{Res}\left(\frac{\Gamma(s)\zeta(2s)t^{-s}}{\zeta(s)}, s = 1/2\right) = \frac{\Gamma(1/2)t^{-1/2}}{2\zeta(1/2)} = \frac{\sqrt{\pi}t^{-1/2}}{2\zeta(1/2)}. \tag{12}$$

Combining the residue contributions (11) and (12) leads to the following heuristic.

Heuristic 4.4 (Liouville Series) *As $r \to 1-$, we have within the computable range*

$$\sum_{n=1}^{\infty}\lambda(n)r^n \text{ `` } \sim \text{ " } 1 + \frac{\sqrt{\pi}}{2\zeta(1/2)}(\log(1/r))^{-1/2} \approx 1 - \frac{0.6068573898}{\sqrt{\log(1/r)}}. \tag{13}$$

The last term in (13), which was not present in the three "Möbius like" series considered above, dominates the behavior of the Liouville power series and explains the larger values observed in Table 2. Table 4 compares these values to those on

Table 4 Numerical values of the Liouville power series along with the heuristic given by (13)

r	$\displaystyle\sum_{n=1}^{\infty} \lambda(n)r^n$	$1 + \dfrac{\sqrt{\pi}}{2\zeta(1/2)\sqrt{\log(1/r)}}$
0.9	−0.872496	−0.8695957
0.99	−5.0533844	−5.0533579
0.999	−18.1857169	−18.1857166
0.9999	−59.6842220	−59.6842218
0.99999	−190.9046778	−190.9046769

the right-hand side of the heuristic (13). As the table shows, the accuracy of the heuristic (13) is uncannily good, and significantly better than in any of the other cases of "Möbius like" functions that we investigated. The main reason for this is that the extra factor $\zeta(2s)$ causes the residue at $s = -1$, which contributed the linear term $12t$ in Fröberg's heuristic (4), to vanish.

5   Concluding Remarks

We conclude with some remarks on related questions and possible applications.

An obvious question is whether heuristics such as Fröberg's formula (4) for the Möbius series can be turned into rigorous statements. In the case of (4) or similar formulas this would seem to be quite difficult, even under strong assumptions such as the Riemann Hypothesis and the simplicity of zeta zeros. Hardy and Littlewood [7, p. 122] make a remark to that effect:

> "Nor does it seem possible, in the present state of knowledge of the properties of $\zeta(s)$, to give a satisfactory proof of the explicit formula for $\sum_1^{\infty} \mu(n)e^{-ny}$."

Hardy and Littlewood contrast this situation with a very similar one, namely the sum $\sum_1^{\infty}(\Lambda(n) - 1)e^{-ny}$, for which they do indeed provide an explicit formula in their paper.

A natural question is to what extent other number-theoretic power series exhibit "fake" convergence of the type observed in the Möbius series. Numerical investigations from [3] suggest that "fake" convergence is extremely rare, although there are many power series that do exhibit some kind of general "fake" asymptotic behavior such as that exhibited by the Liouville series. It seems that almost any type of perturbation of the Möbius function destroys the "fake" convergence property. For example, Möbius series restricted by a congruence condition do not exhibit "fake" convergence, although one can derive more general heuristics for such series from the heuristics on Möbius series with character twists given in Sect. 4. The same holds for Möbius series with exponential twists, i.e., with coefficients $\mu(n)e((a/q)n)$, where $e(t) = e^{2\pi it}$ and a/q is a rational number in $(0, 1)$.

A related question concerns the behavior of the Möbius series $\sum_{n=1}^{\infty} \mu(n)z^n$ as z approaches a point on the unit circle other than 1. Petrushov [9] proved an oscillation result similar to (2) for the case when z approaches a point $e(\alpha)$

on the unit circle with *rational* α. In particular, his result implies that, for such α, the series $\sum_{n=1}^{\infty} \mu(n)e(\alpha n)$ is not Abel summable. For the special case when $z = -e^{-t}$, Fröberg [5, (4.8)] derived a heuristic for $t \to 0$ that is similar to (4), but involves an additional term $(4/\log 2)\log(1/t)$, thus destroying any "fake" convergence behavior. Fröberg also carried out numerical computations of the series with $z = re(\alpha)$, for various other choices of the angle α. His data showed no "fake" convergence behavior when $\alpha \neq 0$. This suggests that the point $z = 1$ is the only point on the unit circle at which the Möbius series exhibits this behavior.

An intriguing question is whether the oscillations predicted in Fröberg's heuristic (4) and (5), i.e., the oscillating part of the graph in Fig. 1, can be "detected" experimentally. In the case of the original Möbius series, these oscillations fall well outside the range in which the series can be computed, as this would require computing the series to around 10^{18} terms. On the other hand, in [3] we did find some series of the form $\sum_{n=1}^{\infty} \mu(n)\chi(n)z^n$, for which it is possible to "see" the oscillations in the experimental data. Specifically, the character needs to be such that the associated L-function has a "low" zero in order to magnify the effect of the oscillating term corresponding to this zero. A systematic search for such characters using the L-function database LMFDB [10] yielded several examples of such characters. The smallest modulus with a noticeable oscillating behavior was 19.

Finally, we remark on a possible application of "fake" asymptotics that was pointed out to the author by Yifan Yang: Namely, one can try to exploit the fact that the "fake" limit of the power series typically involves the value of an underlying Dirichlet series at a special point (such as 0). Thus, computing the power series numerically could provide a way to experimentally determine special values of an underlying L-function. This would be particularly interesting in cases where such values are not known or the subject of conjectures.

Acknowledgements The work described in Sect. 4 was carried out in 2013/2014 by Yiwang Chen, Daniel Hirsbrunner, M. Tip Phaovibul, Dylan Yang, and Tong Zhang as part of an undergraduate research project at the Illinois Geometry Lab [3]. Numerical computations for this work were carried out at the University of Illinois Campus Computing Cluster, a high performance computing platform.

References

1. P.T. Bateman, H. Diamond, On the oscillation theorems of Pringsheim and Landau, in *Number Theory*. Trends in Mathematics (Birkhäuser, Basel, 2000), pp. 43–54
2. J. Bohman, C.-E. Fröberg, Heuristic investigation of chaotic mapping producing fractal objects. BIT **35**, 609–615 (1995)
3. Y. Chen, D. Hirsbrunner, D. Yang, T. Zhang, M. Tip Phaovibul, A.J. Hildebrand, Randomness in number theory: the Möbius case, IGL Project Report 2013/14, https://faculty.math.illinois.edu/~ajh/ugresearch/randomness-spring2014report.pdf
4. H. Delange, Sur certaines séries entières particulières. Acta Arith. **92**, 59–70 (2000)

5. C.-E. Fröberg, Numerical studies of the Möbius power series. Nordisk Tidskr. Informations-Behandling **6**, 191–211 (1966)
6. S. Gerhold, Asymptotic estimates for some number-theoretic power series. Acta Arith. **142**, 187–196 (2010)
7. G.H. Hardy, J.E. Littlewood, Contributions to the theory of the Riemann zeta-function and the theory of the distribution of primes. Acta Math. **41**, 119–196 (1916)
8. I. Katai, On oscillations of number-theoretic functions. Acta Arith. **13**, 107–121 (1967)
9. O. Petrushov, On the behaviour close to the unit circle of the power series with Möbius function coefficients. Acta Arith. **164**, 119–136 (2014)
10. The LMFDB Collaboration, The L-functions and modular forms database (2014), http://www.lmfdb.org

The Convex Hull of the Prime Number Graph

Nathan McNew

Abstract Let p_n denote the n-th prime number, and consider the prime number graph, the collection of points (n, p_n) in the plane. Pomerance uses the points lying on the boundary of the convex hull of this graph to show that there are infinitely many n such that $p_{2n} < p_{n-i} + p_{n+i}$ for all $i < n$. More recently, the primes on the boundary of this convex hull have been considered by Tutaj. We resolve several conjectures of Pomerance and Tutaj by giving improved bounds on the number and distribution of these primes as well as related forms of 'extreme' primes.

1 Introduction

Let p_n denote the n-th prime number, and consider the collection of points (n, p_n) in the plane $\mathbb{R}^2$, which we refer to as the *prime number graph*. The work of Zhang [12] and Maynard [6] and Tao implies there exist infinitely many pairs of consecutive points in this graph with some fixed, finite, integral slope between them. Currently we know this slope is at most 246, and the twin-prime conjecture is true if and only if there are infinitely many such pairs with slope exactly 2. On the other hand, from the prime number theorem, we know that this is not the behaviour of these points on average, as the collection of points which form the prime number graph tends upward faster than any linear function, growing roughly as $n \log n$, so the slopes between the first n consecutive points is about $\log n$ on average.

Because of the irregularities in the gaps between primes, the growth and distribution of these points can be fairly erratic. For example, if one considers the shape that is created by taking all of the line segments between consecutive points in this graph, the result is far from being a convex subset of the plane.

One can take the convex hull of this shape, however, which smooths out most of the irregularity in the primes' distribution, and ask about the collection of points

N. McNew (✉)
Department of Mathematics, Towson University, Towson, MD, USA
e-mail: nmcnew@towson.edu

© Springer International Publishing AG, part of Springer Nature 2018
J. Pintz, M. Th. Rassias (eds.), *Irregularities in the Distribution of Prime Numbers*,
https://doi.org/10.1007/978-3-319-92777-0_7

Fig. 1 The first 30 points on the prime number graph, along with the lower boundary of the convex hull of the prime number graph

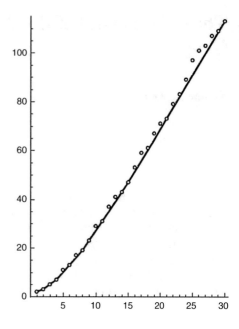

(n, p_n), which are vertex points of this convex hull (Fig. 1). The subset of primes forming such points, which we will refer to in the following as the *convex primes*, was studied by Pomerance in [9], and recently by Tutaj [11] and additionally is discussed in problem A14 of Guy's book of unsolved problems in number theory [5]. In what follows we will let $c_1, c_2, \ldots$ denote the indices of the sequence of convex primes, $p_{c_1}, p_{c_2}, \ldots$.

Pomerance uses the set of convex primes, which he shows is an infinite set, to study a second subset of the primes, the *midpoint convex primes*, which are those prime numbers, p_n, which satisfy the inequality

$$2p_n < p_{n-i} + p_{n+i} \qquad \text{for all positive } i < n. \tag{1}$$

Because this condition is equivalent to the requirement that the line segment connecting any two points $(n - i, p_{n-i})$, and $(n + i, p_{n+i})$ pass above the point (n, p_n) we see that the set of convex primes is clearly a subset of the midpoint convex primes, and so the infinitude of the former set immediately implies the infinitude of the latter.

Pomerance also looks at a multiplicative version of inequality (1), those primes which satisfy the inequality

$$p_n^2 > p_{n-i} p_{n+i} \qquad \text{for all positive } i < n. \tag{2}$$

Primes satisfying this condition have become known as *good primes*. Using the log-prime number graph, the collection of points $(n, \log p_n)$, and specifically those primes, referred to here as the *log-convex primes*, corresponding to the vertices of the convex hull of the log-prime number graph, Pomerance likewise shows that there are infinitely many good primes. This disproved a conjecture of Erdős, who had conjectured that there were only finitely many good primes, while confirming one of Selfridge, who had conjectured the opposite.

Consider the line connecting the origin $(0,0)$ with any vertex of the convex hull, (n, p_n). This line has slope $\frac{p_n}{n}$ and, with the exception of the first few convex primes (specifically the points $(1,2)$, $(2,3)$, and $(4,7)$), one finds that the portion of the line segment to the right of the line $x = 1$ lies entirely within this convex hull, hence strictly above and to the left of its lower boundary. This means that the slopes of these lines must be increasing, and so within the sequence of convex primes p_{c_1}, $p_{c_2}, \ldots$ we have necessarily that

$$\frac{p_{c_i}}{c_i} < \frac{p_{c_{i+1}}}{c_{i+1}} \tag{3}$$

for all $i \geq 2$.

Pomerance points out that a result of Erdős and Prachar [4], which shows that any subsequence of the primes which has the property (3) has relative density 0 in the primes, clearly implies the same for the convex primes, and so the count of the convex primes up to x is $o\left(\frac{x}{\log x}\right)$. Pomerance also notes that another result of Erdős [3] shows that the midpoint convex primes have relative upper density strictly less than one as well.

Note that while the convex primes are closely related to the set of primes p_n with the property that

$$\frac{p_n}{n} < \min_{1 \leq i < \infty} \frac{p_{n+i}}{n+i}, \tag{4}$$

they are not the same set, and in practice it appears that the set of primes satisfying (4) is a substantially larger set, of which the convex primes appear to be a subset.

Tutaj [11] proves the following theorem, conditional on the Riemann Hypothesis.

Theorem 1 (Tutaj) *Assume the Riemann Hypothesis, and let p_{c_i} denote the i-th convex prime. Then*

$$\lim_{i \to \infty} \frac{p_{c_{i+1}}}{p_{c_i}} = 1.$$

He also makes several conjectures, two of which we restate here for reference.

Conjecture 1 (Tutaj) The sum of the reciprocals of the convex primes,

$$\sum_{i=1}^{\infty} \frac{1}{p_{c_i}}$$

converges.

Conjecture 2 (Tutaj) The sum of the reciprocals of the logarithms of the convex primes,

$$\sum_{i=1}^{\infty} \frac{1}{\log p_{c_i}}$$

diverges.

In Theorem 2 we give a substantially better upper bound than $o(\pi(x))$ for the convex primes, namely that their count is $O\left(\pi(x)^{2/3}\right)$. Conjecture 1 follows as a corollary. Both Conjecture 2 and an unconditional version of Theorem 1 follow as corollaries to Theorem 3, in which we prove an upper bound for the size of the gap between two convex primes. (We also get a substantial improvement to this result, Theorem 4, if we assume the Riemann Hypothesis.)

Additionally, we get a lower bound, $e^{\log^{3/5-\epsilon} x}$ for the number of convex primes up to x, and assuming the Riemann Hypothesis this can be improved to $\frac{b' x^{1/4}}{\log^{3/2} x}$ for some constant $b' > 0$. These results were claimed without proof in [9]. In Sect. 4 we confirm another conjecture Pomerance made in that paper, that the log-convex primes have relative density zero in the primes.

Finally, in Sect. 5 we give results of computations on the convex primes and log-convex primes up to 10^{13}. Based on these computations it appears that the exponent $1/4$ arising in the lower bound of the convex primes is likely to be close to the correct power of x in the counting function of the convex primes up to x.

2 Counting the Convex Primes

We give here a substantially improved upper bound for the count of the convex primes, using only the prime number theorem, and the fact that there aren't many possible rational slopes with small denominator. First however, we prove a lemma regarding the slope of a line segment along the edge of the convex hull.

Lemma 1 *If (m, p_m) is any point on the boundary of the convex hull of the prime number graph, then the slope of the line segment of the convex hull following this point has slope*

$$\log m + \log \log m + o(1)$$

as $m \to \infty$.

Proof Suppose, for the sake of contradiction, that there exists arbitrarily large m where (m, p_m) is on the boundary of the convex hull and the slope is greater than $\log m + \log \log m + d$ for some $d > 0$.

Fix $\alpha > 1$ chosen so that $\log \alpha + \frac{1+d}{\alpha} - d \leq 1 - \epsilon$ for some $\epsilon > 0$. Then for sufficiently large m, at those values of m where the slope of the convex hull following p_m is greater than $\log m + \log \log m + d$, we have (using the fact, from the prime number theorem, that $p_m = m(\log m + \log \log m - 1 + o(1)))$ that

$$p_{\lceil \alpha m \rceil} \geq p_m + (\alpha m - m)(\log m + \log \log m + d)$$

$$= m(\log m + \log \log m - 1 + o(1)) + (\alpha m - m)(\log m + \log \log m + d)$$

$$= \alpha m \left(\log \alpha m + \log \log \alpha m - \log \alpha - \frac{1+d}{\alpha} + d + o(1) \right)$$

$$\geq \alpha m (\log \alpha m + \log \log \alpha m - 1 + \epsilon + o(1))$$

as $m \to \infty$, which would contradict the prime number theorem.

Similarly, if we suppose there are arbitrarily large m where the slope is less than $\log m + \log \log m - d$ for some $d > 0$, we can fix a $0 < \beta < 1$ chosen so that $\log \beta + \frac{1-d}{\beta} + d \leq 1 - \epsilon$ for some $\epsilon > 0$. Then for sufficiently large m, where the slope of the convex hull following p_m is less than $\log m + \log \log m - d$, we have that

$$p_{\lceil \beta m \rceil} \geq p_m - (m - \beta m)(\log m + \log \log m - d)$$

$$= m(\log m + \log \log m - 1 + o(1)) - (m - \beta m)(\log m + \log \log m - d)$$

$$= \beta m \left(\log \beta m + \log \log \beta m - \log \beta - \frac{1-d}{\beta} - d + o(1) \right)$$

$$\geq \beta m (\log \beta m + \log \log \beta m - 1 + \epsilon + o(1))$$

as $m \to \infty$, again contradicting the prime number theorem.

We now get an upper bound for the count of the convex primes.

Theorem 2 *The count of the convex primes up to x is $O\left(\frac{x^{2/3}}{\log^{2/3} x}\right)$.*

Proof We count those convex primes in the dyadic interval $(\frac{1}{2}x, x]$. The slopes between the consecutive convex primes are necessarily strictly increasing rational numbers given by

$$\frac{p_{c_{j+1}} - p_{c_j}}{c_{j+1} - c_j}.$$

Because we are counting those $p_{c_j} \in \left(\frac{1}{2}x, x\right]$, we have $\frac{1}{2}\pi(x) < \pi\left(\frac{1}{2}x\right) < c_j \leq \pi(x)$. From Lemma 1 we know that the slope at each p_{c_j} is contained in some interval of length $\log 2 + o(1)$. If $c_j - c_{j-1} = k$, this leaves at most $k(\log 2 + o(1))$ possible values of $p_{c_j} - p_{c_{j-1}}$. Thus, for each integer k, there are at most $O(k)$ pairs $p_{c_j}, p_{c_{j+1}}$ in $(\frac{1}{2}x, x]$ with $c_j - c_{j-1} = k$, or $O(K^2)$ such convex primes which follow a gap between indices of consecutive convex primes at most K apart, a parameter to be chosen shortly.

The number of consecutive convex primes $p_{c_{j-1}}$ and p_{c_j} in the interval $\left(\frac{1}{2}x, x\right]$ with $c_j - c_{j-1} > K$ is at most $O\left(\frac{x}{K \log x}\right)$. Equating K^2 with $\frac{x}{K \log x}$ gives $K = (x/\log x)^{1/3}$ and thus that the total number of convex primes in the interval $(\frac{1}{2}x, x]$ is $O\left(\frac{x^{2/3}}{\log^{2/3} x}\right)$. The result then follows by summing dyadically.

Igor Shparlinski points out that this result also follows from a result of George Andrews [1]. Andrews shows that any convex region of the plane with area A, bounded by line segments whose endpoints lie in $\mathbb{Z}^2$, has at most $O(A^{1/3})$ noncollinear vertices. In this case, the region bounded below by the convex hull of the prime number graph and above by the line segment connecting $(1,2)$ to the vertex corresponding to the greatest convex prime less than x is a convex region with area $O\left(\frac{x^2}{\log x}\right)$.

As mentioned in the introduction, this proves Tutaj's Conjecture 1, showing that the sum of the reciprocals of the convex primes converges. We now give an upper bound for the size of the gap between consecutive convex primes, which will, as a corollary give us a lower bound for the count of the convex primes.

Theorem 3 *There exists a constant $B > 0$ such that for sufficiently large values of i the gap between consecutive convex primes p_{c_i} and $p_{c_{i+1}}$, is bounded above by*

$$p_{c_{i+1}} - p_{c_i} \le p_{c_i} \exp\left\{\frac{-B \log^{3/5} p_{c_i}}{(\log \log p_{c_i})^{1/5}}\right\}.$$

Proof The best known error term for the prime number theorem tells us that

$$c_i = \mathrm{li}(p_{c_i}) + O\left(p_{c_i} \exp\left\{\frac{-A \log^{3/5} p_{c_i}}{(\log \log p_{c_i})^{1/5}}\right\}\right) \tag{5}$$

for some constant $A > 0$. Thus, there exists a constant $D > 0$ such that all of the points of the prime number graph lie between the (implicitly defined as a function of x) curves

$$x = \mathrm{li}(y) - Dy \exp\left\{\frac{-A \log^{3/5} y}{(\log \log y)^{1/5}}\right\} \tag{6}$$

and

$$x = \mathrm{li}(y) + Dy \exp\left\{\frac{-A \log^{3/5} y}{(\log \log y)^{1/5}}\right\}. \tag{7}$$

We will find the maximal length of a line segment bounded between these two curves, and thus the maximal length of a line segment forming part of the boundary of the convex hull of the prime number graph.

Because both of these curves are convex and curved upward with unbounded derivative, we see that any line partially lying between these two curves with a sufficiently steep, positive slope will intersect the outer curve (7) twice. Since the slopes of the lines forming the boundary of the convex hull tend to infinity and the line segments necessarily exist between these two curves such a segment must lie below and to the right of the inner curve (6). In fact the longest that such a line segment could possibly be, given these constraints, would be if it began and ended at the boundary of the outer curve and was tangent to the inner curve in between.

Suppose then that a line segment contained between these two curves intersects (7) at points (x_1, y_1) and (x_2, y_2), $x_1 < x_2$. Let $\Delta = y_2 - y_1$ and let $(x_0, y_0) = \left(\frac{x_1+x_2}{2}, \frac{y_1+y_2}{2}\right)$ be the midpoint of this line.

In order for our line segment to remain to the right of (6), we need in particular that the midpoint of this line segment does, and so

$$\text{li } y_0 - D y_0 \exp\left\{\frac{-A \log^{3/5} y_0}{(\log\log y_0)^{1/5}}\right\} < x_0 = \frac{x_1 + x_2}{2}$$

$$= \frac{\text{li } y_1}{2} + \frac{D}{2} y_1 \exp\left\{\frac{-A \log^{3/5} y_1}{(\log\log y_1)^{1/5}}\right\}$$

$$+ \frac{\text{li } y_2}{2} + \frac{D}{2} y_2 \exp\left\{\frac{-A \log^{3/5} y_2}{(\log\log y_2)^{1/5}}\right\}. \tag{8}$$

Using the Taylor series for $\text{li } y$, we have that

$$\text{li }(y + t) = \text{li } y + \frac{t}{\log y} - \frac{t^2}{2y \log^2 y} + O\left(\frac{t^3}{y^2 \log^2 y}\right) \tag{9}$$

and so

$$\text{li } y_0 - \frac{\text{li } y_1 + \text{li } y_2}{2} = \text{li } y_1 + \frac{\Delta}{2} - \frac{\text{li } y_1 + \text{li } y_1 + \Delta}{2}$$

$$= \text{li } y_1 + \frac{\Delta}{2\log y_1} - \frac{\Delta^2}{8y_1 \log^2 y_1} - \text{li } y_1 - \frac{\Delta}{2\log y_1}$$

$$+ \frac{\Delta^2}{4y_1 \log^2 y_1} + O\left(\frac{\Delta^3}{y_1^2 \log^2 y_1}\right)$$

$$= \frac{\Delta^2}{8y_1 \log^2 y_1} + O\left(\frac{\Delta^3}{y_1^2 \log^2 y_1}\right). \tag{10}$$

Using this in (9), we get that

$$\frac{\Delta^2}{8y_1 \log^2 y_1} + O\left(\frac{\Delta^3}{y_1^2 \log^2 y_1}\right)$$

$$< D\left(y_0 \exp\left\{\frac{-A \log^{3/5} y_0}{(\log\log y_0)^{1/5}}\right\} + \frac{y_1}{2} \exp\left\{\frac{-A \log^{3/5} y_1}{(\log\log y_1)^{1/5}}\right\}\right.$$

$$\left. + \frac{y_2}{2} \exp\left\{\frac{-A \log^{3/5} y_2}{(\log\log y_2)^{1/5}}\right\}\right)$$

$$< D \exp\left\{\frac{-A \log^{3/5} y_1}{(\log\log y_1)^{1/5}}\right\}\left(y_0 + \frac{y_1}{2} + \frac{y_2}{2}\right)$$

$$= D \exp\left\{\frac{-A \log^{3/5} y_1}{(\log\log y_1)^{1/5}}\right\}(2y_1 + \Delta), \tag{11}$$

so that

$$\Delta^2 < Dy_1 \log^2 y_1 \exp\left\{\frac{-A \log^{3/5} y_1}{(\log\log y_1)^{1/5}}\right\}(2y_1 + \Delta) + O\left(\frac{\Delta^3}{y_1}\right), \tag{12}$$

which means that

$$\Delta \ll y_1 \log y_1 \exp\left\{\frac{-A \log^{3/5} y_1}{2(\log\log y_1)^{1/5}}\right\}. \tag{13}$$

This upper bound for Δ means that the gap between the convex primes p_{c_i} and $p_{c_{i+1}}$ is

$$p_{c_{i+1}} - p_{c_i} \ll p_{c_i} \log p_{c_i} \exp\left\{\frac{-A \log^{3/5} p_{c_i}}{2(\log\log p_{c_i})^{1/5}}\right\}$$

$$\leq p_{c_i} \exp\left\{\frac{-B \log^{3/5} p_{c_i}}{(\log\log p_{c_i})^{1/5}}\right\}.$$

for some constant $B > 0$.

Note that if we assume the Riemann Hypothesis, we can use the stronger form of the prime number theorem, that

$$n = \mathrm{li}\, p_n + O\left(\sqrt{p_n} \log p_n\right),$$

and the same proof gives the following improvement.

Theorem 4 *Assume the Riemann Hypothesis. Then the gap between consecutive convex primes p_{c_i} and $p_{c_{i+1}}$, is bounded above by*

$$p_{c_{i+1}} - p_{c_i} \ll p_{c_i}^{3/4} \log^{3/2} p_{c_i}.$$

It was claimed without proof in [9] that the best known results on the error term in the prime number theorem imply that the count of the convex primes up to x is at least $e^{c \log^{3/5-\epsilon} x}$ for some $c > 0$. This follows as an immediate corollary to Theorem 3.

Corollary 1 *There exists a constant $B > 0$ such that the count of the number of convex primes up to x is at least*

$$\exp \left\{ \frac{B \log^{3/5} x}{(\log \log x)^{1/5}} \right\}.$$

Likewise, if one assumes the Riemann Hypothesis, then we get the following as a corollary to Theorem 4.

Corollary 2 *Assume the Riemann Hypothesis. Then there exists a constant $B' > 0$ such that the count of the number of convex primes up to x is at least*

$$\frac{B' x^{1/4}}{\log^{3/2} x}.$$

The lower bound of Corollary 1 also has another corollary which proves Conjecture 2.

Corollary 3 *The sum of the reciprocals of the logarithms of the convex primes,*

$$\sum_{i=1}^{\infty} \frac{1}{\log p_{c_i}}$$

diverges.

Furthermore, by showing that $p_{c_{i+1}} - p_{c_i} = o(p_{c_i})$, Theorem 3 gives an unconditional proof of Theorem 1, that the ratio of consecutive convex primes converges to 1.

Corollary 4 *Let p_{c_i} denote the i-th convex prime. Then*

$$\lim_{i \to \infty} \frac{p_{c_{i+1}}}{p_{c_i}} = 1.$$

3 Edge Convex Primes

The results discussed above, except for Lemma 1, apply only for points (n, p_n) which are vertex points of the convex hull of the prime number graph. It is possible for points to lie on the boundary of the convex hull without being vertex points, for example, the point $(3, 5)$ lies on the line segment between the vertex points $(2, 3)$ and $(4, 7)$.

In the computation of the convex primes up to 10^{13} (discussed further in Sect. 5) we find only five examples of primes with this property, namely $5, 13, 23, 31$, and 43. Based on this evidence we conjecture the following.

Conjecture 3 There are only finitely many primes p_n for which the point (n, p_n) lies on the boundary of the convex hull of the prime number graph without being a vertex point of it.

Despite this conjecture, obtaining upper bounds for the number of these *edge convex primes* is more difficult than for the convex primes. Combining the results of Theorem 3 with the ideas of Theorem 2 we can get the following upper bound.

Theorem 5 *The count of the edge convex primes, those primes which lie on the boundary of the convex hull of the prime number graph without being vertices of it, is $O\left(x \exp\left\{-b' \frac{\log^{3/5} x}{(\log\log x)^{1/5}}\right\}\right)$ for some constant $b' > 0$.*

Proof We count those edge convex primes in the dyadic interval $(\frac{1}{2}x, x]$. As noted before, the slopes of the lines forming the boundary of the convex hull are necessarily strictly increasing rational numbers contained in some interval of length $\log 2 + o(1)$, from Lemma 1. From Theorem 3 we see that there are necessarily at least $\exp\left\{B\frac{\log^{3/5} x}{(\log\log x)^{1/5}}\right\}$ distinct line segments in this interval (for some $B > 0$), and that each line segment ranges over $O\left(x \exp\left\{-B\frac{\log^{3/5} x}{(\log\log x)^{1/5}}\right\}\right)$ primes.

If a line segment of the boundary of the convex hull has slope $\frac{a}{d}$, with a and d coprime, then the edge convex primes lying on this line segment must be spaced at least d primes apart, so the maximum number of edge convex primes contained on such an interval is $O\left(\frac{x}{d} \exp\left\{-B\frac{\log^{3/5} x}{(\log\log x)^{1/5}}\right\}\right)$.

Since each of the slopes is a rational number contained in an interval of length $\log 2 + o(1)$ there are $O(\varphi(d))$ line segments forming the convex hull in this range with slope whose denominator in lowest terms is d, and thus the total number of edge convex primes that can lie on any line segment whose slope has denominator at most D is $O\left(Dx \exp\left\{-B\frac{\log^{3/5} x}{(\log\log x)^{1/5}}\right\}\right)$. Because the total number of edge convex primes which lie on those line segments whose slope has denominator (in reduced form) greater than D is $O\left(\frac{x}{D}\right)$, we find, by optimizing D, that the total number of edge convex primes in this interval is $O\left(x \exp\left\{-\frac{1}{2}B\frac{\log^{3/5} x}{(\log\log x)^{1/5}}\right\}\right)$. The result then follows by summing dyadically.

Assuming the Riemann Hypothesis, this can be improved using Theorem 4.

Theorem 6 *Assuming the Riemann Hypothesis, we have that the count of the edge convex primes is $O\left(x^{7/8} \log^{3/4} x\right)$.*

4   Multiplicatively Convex Primes

In the paper [9] where Pomerance introduced the convex primes, he also considered the log-convex primes, those primes forming the vertices of the boundary of the convex hull of the log-prime number graph. Unlike the convex primes, he was not able to show that they had relative density zero among the primes, but conjectured that this was the case. We prove that this is in fact the case here. First however we prove a lemma regarding the slope of the boundary of the convex hull of the log-prime number graph.

In the following we adopt the notation of [2] and use ali(x) to denote the functional inverse of the logarithmic integral function, li(x).

Lemma 2 *If $(m, \log p_m)$ is any point on the boundary of the convex hull of the log-prime number graph, then the slope of the line segment of the convex hull following this point has slope*

$$\frac{\text{ali}'(m)}{\text{ali}(m)} + O\left(\frac{1}{m}\exp\left\{-\frac{A\log^{3/5}m}{2(\log\log m)^{1/5}}\right\}\right) \tag{14}$$

as $m \to \infty$, for some $A > 0$.

Proof We can rewrite Eq. (5), the strongest known form of the prime number theorem, as

$$p_n = \text{ali}(n) + O\left(n\exp\left\{-A\frac{\log^{3/5}n}{(\log\log n)^{1/5}}\right\}\right), \tag{15}$$

and so,

$$\log p_n = \log \text{ali}(n) + O\left(\exp\left\{-A\frac{\log^{3/5}n}{(\log\log n)^{1/5}}\right\}\right). \tag{16}$$

Note that

$$\frac{d}{dx}\log \text{ali}(x) = \frac{\text{ali}'(x)}{\text{ali}(x)} \sim \frac{\log x}{x\log x} = \frac{1}{x}$$

and that

$$\frac{d^2}{dx^2}\log \text{ali}(x) \sim -\frac{1}{x^2}.$$

Suppose rather, for contradiction, that for any arbitrarily large constant C, there exist infinitely many d_i where the slope following the point $(d_i, \log p_{d_i})$ is less than $\frac{\text{ali}'(d_i)}{\text{ali}(d_i)} - \frac{C}{d_i} \exp\left\{ -\frac{A \log^{3/5} d_i}{2(\log \log d_i)^{1/5}} \right\}$.

Let $f(x) = \exp\left\{ -\frac{A \log^{3/5} x}{2(\log \log x)^{1/5}} \right\}$. Because the slopes of the boundary of the convex hull of the log-prime number graph are strictly decreasing, we see that for such values of d_i we have, using the Taylor series approximation of $\log \text{ali}(x)$, that

$$
\begin{aligned}
\log p_{\lfloor d_i + d_i f(d_i) \rfloor} &\leq \log p_{d_i} + d_i f(d_i) \left(\frac{\text{ali}'(d_i)}{\text{ali}(d_i)} - \frac{C}{d_i} f(d_i) \right) \\
&= \log \text{ali}(d_i) + d_i f(d_i) \frac{\text{ali}'(d_i)}{\text{ali}(d_i} - C f(d_i)^2 + O\left(f(d_i)^2 \right) \\
&= \log \text{ali}\, (d_i + d_i f(d_i)) - C f(d_i)^2 + O\left(f(d_i)^2 \right). \quad (17)
\end{aligned}
$$

Because the constant C can be taken arbitrarily large, Eq. (17) contradicts Eq. (16), thus proving that the slope is at least as large as expression (14). The corresponding upper bound follows by essentially the same argument.

We are now able to see that the log-convex primes have relative density zero among the primes.

Theorem 7 *The count of the number of log-convex primes up to x, as $x \to \infty$, is at most $\dfrac{x}{\log^{4/3 - o(1)} x}$.*

Proof Denote by $D(t, x)$ the number of primes $p \leq x$ for which $p + t$ is also prime. Brun's sieve can be used to show (see, for example, [10]) that

$$
D(t, x) \leq c \left(\prod_{\substack{p \mid t \\ p \leq x}} \left(1 - \frac{1}{p} \right)^{-1} \right) \frac{x}{\log^2 x}
$$

for some absolute constant c.

Fix x and consider pairs of primes

$$
\frac{x}{\log x} \leq p_n < p_{n+k} \leq x \quad (18)
$$

with $k \leq \log^{1/3} x$. (The number of log-convex primes less than $\frac{x}{\log x}$ is already at most $\pi\left(\frac{x}{\log x} \right) \sim \frac{x}{\log^2 x}$ so their contribution is insignificant.)

Suppose a pair of primes p_n and p_{n+k} in this range are consecutive log-convex primes. Then the slope of the line connecting the corresponding points is $\frac{\log p_{n+k} - \log p_n}{k}$, and so Lemma 2 tells us that

$$\left| \frac{\log p_{n+k} - \log p_n}{k} - \frac{\mathrm{ali}'(n)}{\mathrm{ali}\, n} \right| = O\left(\frac{1}{n} \exp\left\{ -A \frac{\log^{3/5} n}{(\log\log n)^{1/5}} \right\} \right)$$

for some $A > 0$. Letting $\Delta = p_{n+k} - p_n$, we can write

$$\frac{\log p_{n+k} - \log p_n}{k} = \frac{\log(p_n + \Delta) - \log p_n}{k} = \frac{\Delta}{kp_n} + O\left(\frac{\Delta^2}{kp_n^2} \right). \qquad (19)$$

So, assuming $\Delta \leq n^\epsilon$, and using the fact that

$$p_n \frac{\mathrm{ali}'(n)}{\mathrm{ali}\, n} = \log n + O(\log\log n) = \log x + O(\log\log x)$$

for n in the range (18), we find that

$$|\Delta - k \log x| \leq kb \log\log x \qquad (20)$$

for some absolute constant b.

Thus, the total number of such pairs is bounded above by

$$\sum_{1 \leq k \leq \log^{1/3} x} \sum_{t \in \mathcal{I}_k} D(t, x) \leq \sum_{1 \leq k \leq \log^{1/3} x} \sum_{t \in \mathcal{I}_k} c \left(\prod_{p|t} \left(1 - \frac{1}{p}\right)^{-1} \right) \frac{x}{\log^2 x} \qquad (21)$$

where $\mathcal{I}_k = \{t : |t - k \log x| \leq bk \log\log x\}$. Now, using Mertens' theorem,

$$\sum_{t \in \mathcal{I}_k} \prod_{p|t} \left(1 - \frac{1}{p}\right)^{-1} < 2bk \log\log x \prod_{p \leq k(\log x + b \log\log x)} \left(1 - \frac{1}{p}\right)^{-1}$$

$$= O(k \log\log x \log(k \log x))$$

$$= O(k(\log\log x)^2).$$

So,

$$\sum_{1 \leq k \leq \log^{1/3} x} \sum_{t \in \mathcal{I}_k} D(t, x) \ll \sum_{1 \leq k \leq \log^{1/3} x} k(\log\log x)^2 \frac{x}{\log^2 x} < (\log\log x)^2 \frac{x}{\log^{4/3} x}. \qquad (22)$$

Any log-convex primes not involved in one of these pairs must be spaced at least $\log^{1/3} x$ primes distant from the next log-convex prime, so the count of such isolated log-convex primes is less than $\frac{1}{\log^{1/3} x} \pi(x)$. Thus the contribution from (22) dominates, and we see that the count of the log-convex primes up to x is at most $\frac{x}{\log^{4/3 - o(1)} x}$.

5   Data and Future Work

The table below gives the count $C(x)$ of the number of convex primes up to x for x ranging up to 10^{13}, as well as the exponent one could raise x to in order to approximate this value of $C(x)$. Based on the data, it appears that the lower bound on the order of $\frac{x^{1/4}}{\log^{3/2} x}$, (obtained by assuming the Riemann Hypothesis), Corollary 2 may not be far from the truth, though it seems that $C(x)$ may be growing like x^c for some constant c closer to 0.285.

x	$C(x) = \#$ of convex primes up to x	$\dfrac{\log C(x)}{\log x}$
10^1	3	0.47712
10^2	6	0.38908
10^3	12	0.35973
10^4	22	0.33561
10^5	36	0.31126
10^6	65	0.30215
10^7	121	0.29754
10^8	223	0.29354
10^9	413	0.29066
10^{10}	756	0.28785
10^{11}	1409	0.28626
10^{12}	2621	0.28487
10^{13}	5150	0.28552

As mentioned in Sect. 3, there were only five edge convex primes in this range, all at most 43. From these data it seems natural to conjecture that there are only finitely many edge convex primes, even though the upper bound we have for their count, Theorem 5, is far weaker than our bound for the convex primes, Theorem 2.

Pomerance originally studied the convex primes as a way to show that there were infinitely many midpoint convex primes, those primes satisfying (3), since the convex primes are a subset of the midpoint convex primes. The midpoint convex primes appear to be far more numerous than the convex primes.

On the other hand, it seems clear from the data that the midpoint convex primes are a relatively sparse subset of the primes, growing slightly faster than $\sqrt{x}$. It remains open, however, to prove even that their count is $o(\pi(x))$.

In light of (1), the midpoint convex primes can be characterized as those primes p_n, where the quantity

$$M_n = \min_{1 \le i < n} (p_{n+i} + p_{n-i}) - 2p_n \qquad (23)$$

x	$M(x) =$ # of midpoint convex primes up to x	$\dfrac{\log M(x)}{\log x}$
10^1	3	0.47712
10^2	8	0.45154
10^3	25	0.46598
10^4	89	0.48735
10^5	288	0.49188
10^6	1148	0.50999
10^7	4504	0.52194
10^8	17,293	0.52973
10^9	71,804	0.53957
10^{10}	283,737	0.54529
10^{11}	1,195,764	0.55251

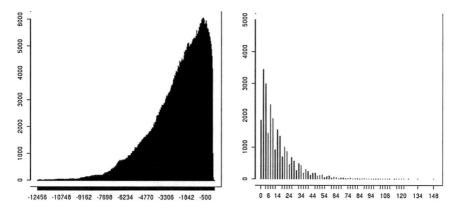

Fig. 2 Distribution of the minimum of M_n for $n < 1.6 \times 10^8$ and the non-negative part of that distribution

is positive. Included in Fig. 2 are two histograms, showing the distribution of M_n for $n < 1.6 \times 10^8$. The first shows the entire distribution, the second shows the (very minor) tail of the distribution for non-negative values.

The distribution of this quantity M_n would be interesting to study further. Based on the data it appears likely that M_n can be arbitrarily large. It can easily be seen that M_n can be arbitrarily negative, as even

$$p_{n+1} + p_{n-1} - 2p_n = (p_{n+1} - p_n) - (p_n - p_{n-1}),$$

the difference between consecutive gaps, can be arbitrarily negative. In fact Pintz [8] shows that $\liminf\limits_{n \to \infty} \frac{p_{n+1} - p_n}{p_{n-1} - p_n} = 0$. Note also how the values of M_n have a tendency to avoid multiples of 6.

We also give the corresponding counts for the log-convex primes and the good primes, which appear to be somewhat more numerous than the convex primes and midpoint convex primes, respectively.

x	$L(x) =$# of log-convex primes up to x	$\dfrac{\log L(x)}{\log x}$	$G(x) = $ # of good primes up to x	$\dfrac{\log G(x)}{\log x}$
10^1	1	0	1	0
10^2	9	0.47712	11	0.52070
10^3	25	0.46598	44	0.54782
10^4	56	0.43705	176	0.56138
10^5	111	0.40906	671	0.56534
10^6	248	0.39908	2668	0.57103
10^7	533	0.38953	10,942	0.57701
10^8	1060	0.37816	45,150	0.58183
10^9	2182	0.37098	189,365	0.58637
10^{10}	4555	0.36585		
10^{11}	9394	0.36117		
10^{12}	19,510	0.35752		
10^{13}	40,901	0.35475		

It would be interesting to develop a heuristic argument, possibly using a probabilistic model for the prime numbers, in order to conjecture what the correct asymptotic growth rate of these counts might be. Obviously, proving asymptotic formulae for these counts would be ideal, though this may be difficult. It would be interesting if progress could be made on even the following questions:

Question 1 Can one prove that the counting function of the midpoint convex primes $M(x)$ is $o(\pi(x))$? Or, likewise, that the counting function for the good primes $G(x)$ is $o(\pi(x))$?

Question 2 Clearly $C(x) < M(x)$ and $L(x) < G(x)$. Can we prove that $C(x) = o(M(x))$, or that $L(x) = o(G(x))$?

One could also consider the convex hull of other graphs related to the prime numbers to study other sorts of "extreme primes." For example, the collection of points $(n, \text{li}(p_n))$ will have infinitely many points on its convex hull both above and below. (To see this, note that the slopes of the bounding convex hull approach 1, but can never reach 1, since both $\limsup\limits_{n\to\infty}(n - \text{li}(p_n)) = \infty$ and $\liminf\limits_{n\to\infty}(n - \text{li}(p_n)) = -\infty$.)

Acknowledgements A portion of this work appeared in the author's Ph.D. thesis [7], written under the direction of Carl Pomerance, whose helpful suggestions were invaluable in the development of this paper. The author is also grateful to Igor Shparlinski, Robin Pemantle, and the anonymous referee for useful suggestions.

References

1. G. Andrews, A lower bound for the volume of strictly convex bodies with many boundary lattice points. Trans. Am. Math. Soc. **106**, 270–279 (1963). MR 0143105 (26 #670)
2. J.A. de Reyna, J. Toulisse, The n-th prime asymptotically. J. Théor. Nombres Bordeaux **25**(3), 521–555 (2013). MR 3179675
3. P. Erdős, On the difference of consecutive primes. Bull. Am. Math. Soc. **54**, 885–889 (1948). MR 0027009 (10,235b)
4. P. Erdős, K. Prachar, Sätze und Probleme über p_k/k. Abh. Math. Sem. Univ. Hamburg **25**, 251–256 (1961/1962). MR 0140481 (25 #3901)
5. R. Guy, *Unsolved Problems in Number Theory* (Springer, New York, 2004)
6. J. Maynard, Small gaps between primes. Ann. Math. (2) **181**(1), 383–413 (2015). MR 3272929
7. N. McNew, Multiplicative problems in combinatorial number theory, Ph.D. thesis, Dartmouth College, 2015
8. J. Pintz, Polignac numbers, conjectures of Erdős on gaps between primes, arithmetic progressions in primes, and the bounded gap conjecture, in *From Arithmetic to Zeta-Functions* (Springer, Cham, 2016), pp. 367–384. MR 3644045
9. C. Pomerance, The prime number graph. Math. Comput. **33**(145), 399–408 (1979)
10. C. Pomerance, A. Sárközy, *Combinatorial Number Theory*. Handbook of Combinatorics, vols. 1, 2 (Elsevier, Amsterdam, 1995), pp. 967–1018. MR 1373676 (97e:11032)
11. E. Tutaj, Prime numbers with a certain extremal type property (2014). Preprint. arXiv:1408.3609
12. Y. Zhang, Bounded gaps between primes. Ann. Math. (2) **179**(3), 1121–1174 (2014). MR 3171761

Irregular Behaviour of Class Numbers and Euler-Kronecker Constants of Cyclotomic Fields: The Log Log Log Devil at Play

Pieter Moree

Abstract Kummer (1851) and, many years later, Ihara (2005) both posed conjectures on invariants related to the cyclotomic field $\mathbb{Q}(\zeta_q)$ with q a prime. Kummer's conjecture concerns the asymptotic behaviour of the first factor of the class number of $\mathbb{Q}(\zeta_q)$ and Ihara's the positivity of the Euler-Kronecker constant of $\mathbb{Q}(\zeta_q)$ (the ratio of the constant and the residue of the Laurent series of the Dedekind zeta function $\zeta_{\mathbb{Q}(\zeta_q)}(s)$ at $s = 1$). If certain standard conjectures in analytic number theory hold true, then one can show that both conjectures are true for a set of primes of natural density 1, but false in general. Responsible for this are irregularities in the distribution of the primes.

With this survey we hope to convince the reader that the apparently dissimilar mathematical objects studied by Kummer and Ihara actually display a very similar behaviour.

1 Introduction

Making conjectures in analytic prime number theory is a notoriously dangerous endeavour,[1] certainly if the basis for this is mostly of numerical nature. The danger lies in the fact that computers can barely spot log log terms and are certainly blind to the log log log terms that frequently occur. The presence of such terms can result in the conjecture being false on very thin subsequences. Celebrated examples are the $\pi(x) < \text{Li}(x)$ conjecture and the Mertens conjecture that $|\sum_{n \leq x} \mu(n)| < \sqrt{x}$ for $n \geq 1$ (for notation see Sect. 2.1). Both of them are false, but true up to gigantic

---

[1]In fact, the title of this paper ends with a question mark. Since it is considered very bad style to have it in the title of a paper, this footnote might be a better place. Not putting the question mark would go against the moral of this paper.

P. Moree (✉)
Max Planck Institute for Mathematics, Bonn, Germany
e-mail: moree@mpim-bonn.mpg.de

© Springer International Publishing AG, part of Springer Nature 2018
J. Pintz, M. Th. Rassias (eds.), *Irregularities in the Distribution of Prime Numbers*,
https://doi.org/10.1007/978-3-319-92777-0_8

values of x. A way out of the danger zone is to change "for all" to some slightly weaker statement. However, this requires a substantial theoretical insight into the conjecture.

Here we present two further conjectures (due to Kummer and Ihara, respectively) where the phenomena indicated above also seem to arise. The final verdict on them is still open but, assuming some standard conjectures from analytic number theory, they are false on some very thin sequences of primes due to irregularities in the distribution of the primes. At a first glance the two conjectures look unrelated. However, they are both connected with the distribution of special L-values and the results and conjectures we present on them are strikingly similar.[2]

In the remaining part of the introduction we formulate the conjectures (after stating some background material) and discuss how they are related to special L-values. In the rest of the paper we discuss results and related conjectures.

Although results from various papers are mentioned in this survey, our main inspiration are Ford et al. [7] for the Euler-Kronecker constant and Granville [11] for Kummer's conjecture. Euler-Kronecker constants for non-quadratic fields were put on the mathematical map mainly thanks to the efforts of Ihara [15–17].

1.1 The Euler-Kronecker Constant for Number Fields

For a number field K we can define, for $\operatorname{Re} s > 1$, the *Dedekind zeta function* by

$$\zeta_K(s) = \sum_{\mathfrak{a}} \frac{1}{N\mathfrak{a}^s} = \prod_{\mathfrak{p}} \frac{1}{1 - N\mathfrak{p}^{-s}}.$$

Here, $\mathfrak{a}$ runs over the non-zero ideals in $\mathcal{O}_K$, the ring of integers of K, $\mathfrak{p}$ runs over the prime ideals in $\mathcal{O}_K$ and $N\mathfrak{a}$ is the norm of $\mathfrak{a}$. It is known that $\zeta_K(s)$ can be analytically continued to $\mathbb{C} - \{1\}$, and that at $s = 1$ it has a simple pole and residue α_K. The prime ideals having prime norm are of particular importance as they are the cause for this pole.

After a suitable normalization with gamma factors, one obtains from $\zeta_K(s)$ a function $\tilde{\zeta}_K(s)$ satisfying the functional equation

$$\tilde{\zeta}_K(s) = \tilde{\zeta}_K(1 - s).$$

Since $\tilde{\zeta}_K(s)$ is entire of order 1, one has the following Hadamard product factorization:

$$\tilde{\zeta}_K(s) = \tilde{\zeta}_K(0)e^{\beta_K s} \prod_{\rho} \left(1 - \frac{s}{\rho}\right) e^{\frac{s}{\rho}}, \tag{1}$$

with $\beta_K \in \mathbb{C}$ and where ρ runs over the zeros of $\zeta_K(s)$ in the critical strip.

---

[2]The similarity was first noted by Andrew Granville, see acknowledgment.

Around $s = 1$ we have the Laurent expansion

$$\zeta_K(s) = \frac{\alpha_K}{s-1} + c_K + c_1(K)(s-1) + c_2(K)(s-1)^2 + \dots. \tag{2}$$

The constant $\gamma_K = c_K/\alpha_K$ is called the *Euler-Kronecker constant* in Ihara [15] and Tsfasman [35]. In particular, we have $c_{\mathbb{Q}} = \gamma = 0.57721566\dots$, the *Euler-Mascheroni constant*, see, e.g., Lagarias [20] for a wonderful survey of related material. In case K is imaginary quadratic, the well-known Kronecker limit formula expresses γ_K in terms of special values of the Dedekind η-function.

An alternative formula for γ_K is given by

$$\gamma_K = \lim_{s\downarrow 1}\left(\frac{\zeta_K'(s)}{\zeta_K(s)} + \frac{1}{s-1}\right), \tag{3}$$

which shows that γ_K is the constant part in the Laurent series of the logarithmic derivative of $\zeta_K(s)$. Using the Hadamard factorization (1) one can relate γ_K to the sum of the reciprocal zeros of $\zeta_K(s)$, cf. [7, p. 1452]. Indeed, in a lot of the literature the logarithmic derivative of the right-hand side of (1) is the starting point in studying γ_K. The main tool of Ihara, cf. [15, p. 411], is an "explicit" formula for the prime function

$$\Phi_K(x) = \frac{1}{x-1} \sum_{N\mathfrak{p}^k \le x} \left(\frac{x}{N\mathfrak{p}^k} - 1\right) \log N\mathfrak{p}, \quad x > 1,$$

relating it to the zeros of $\zeta_K(s)$.

Given any Dirichlet series $L(s)$ with a pole at $s = 1$, we can define its Euler-Kronecker constant as the constant part in the Laurent series of its logarithmic derivative (if this constant exists). In Moree [25] this is considered in case when S is a multiplicative set of integers (that is, for coprime integers m and n one has $mn \in S$ if and only if both m and n are in S) and $L_S(s) = \sum_{n\in S} n^{-s}$ is its associated Dirichlet series.

Another alternative formula for γ_K is given by

$$\gamma_K = \lim_{x\to\infty}\left(\log x - \sum_{N\mathfrak{p}\le x} \frac{\log N\mathfrak{p}}{N\mathfrak{p} - 1}\right). \tag{4}$$

This result is due to de la Vallée-Poussin (1896) in case $K = \mathbb{Q}$ and can be easily generalized to other number fields and settings, cf. [12, 13].

Ihara [15, Theorem 1 and Proposition 3] proved that GRH (Conjecture 6 below) implies that there are absolute constants $c_1, c_2 > 0$ such that

$$-c_1 \log |d_K| < \gamma_K < c_2 \log\log |d_K|, \tag{5}$$

where d_K denotes the discriminant $K/\mathbb{Q}$. Tsfasman [35] showed that the above lower bound is sharp, namely, assuming GRH he proved that

$$\liminf \frac{\gamma_K}{\log |d_K|} \geq -0.13024\ldots,$$

where we range over the number fields K with $|d_K| \to \infty$. Later Badzyan [3] proved that one can take $c_1 = (1 - 1/\sqrt{5})/2 \approx 0.276393$. It is an open problem whether this is sharp.

1.2 The Euler-Kronecker Constant for Cyclotomic Fields

It is a natural question to ask how the Euler-Kronecker constant varies over families of number fields such as quadratic fields and (maximal) cyclotomic fields. After quadratic fields, cyclotomic fields have been most intensively studied (see [21, 36] for book length treatments). Many of the associated quantities of a cyclotomic field $\mathbb{Q}(\zeta_m)$ are explicitly known. Relevant examples for us are their ring of integers, $\mathbb{Z}[\zeta_m]$, and their discriminant

$$d_{\mathbb{Q}(\zeta_n)} = (-1)^{\varphi(n)/2} n^{\varphi(n)} \prod_{p|n} p^{-\frac{\varphi(n)}{(p-1)}}. \tag{6}$$

Moreover, the splitting of a rational prime p into prime ideals in $\mathbb{Z}[\zeta_m]$ of a cyclotomic field follows an easy pattern, see, e.g., [31, Theorem 4.16], which we recall here.

For p a prime not dividing the integer m, we define $\mathrm{ord}_p(m)$ to be the (multiplicative) order of p in $(\mathbb{Z}/m\mathbb{Z})^*$

Lemma 1 (Cyclotomic Reciprocity Law) *Let $K = \mathbb{Q}(\zeta_m)$. If the prime p does not divide m and $f = \mathrm{ord}_p(m)$, then the principal ideal $p\mathcal{O}_K$ factorizes as $\mathfrak{p}_1 \cdots \mathfrak{p}_g$ with $g = \varphi(m)/f$ and all $\mathfrak{p}_i$ are distinct and of degree f.*

However, if p divides m, $m = p^a m_1$ with $p \nmid m_1$ and $f = \mathrm{ord}_p(m_1)$, then $p\mathcal{O}_K = (\mathfrak{p}_1 \cdots \mathfrak{p}_g)^e$ with $e = \varphi(p^a)$, $g = \varphi(m_1)/f$ and all $\mathfrak{p}_i$ are distinct and of degree f.

For notational convenience we will write γ_m instead of $\gamma_{\mathbb{Q}(\zeta_m)}$. Our main focus is on the case where $m = q$ is a prime (unless specified otherwise, m denotes a positive integer and p and q primes). Then we have

$$\zeta_{\mathbb{Q}(\zeta_q)}(s) = \zeta(s) \prod_{\chi \neq \chi_0} L(s, \chi), \tag{7}$$

where χ ranges over the non-trivial characters modulo q, leading to

$$\gamma_q = \gamma + \sum_{\chi \neq \chi_0} \frac{L'(1, \chi)}{L(1, \chi)}. \tag{8}$$

Thus the behaviour of γ_q is related to that of $L(s, \chi)$ and $L'(s, \chi)$ at $s = 1$. (Here and in the rest of the paper we often use the fundamental fact that $L(1, \chi) \neq 0$.)

Let $\mathbb{Q}(\zeta_m)^+$ denote the maximal real subfield of $\mathbb{Q}(\zeta_m)$ and γ_m^+ its Euler-Kronecker constant. In that case we find

$$\zeta_{\mathbb{Q}(\zeta_q)^+}(s) = \zeta(s) \prod_{\substack{\chi \neq \chi_0 \\ \chi(-1)=1}} L(s, \chi). \tag{9}$$

Logarithmic differentiation of the latter product identity then yields

$$\gamma_q^+ = \gamma + \sum_{\substack{\chi \neq \chi_0 \\ \chi(-1)=1}} \frac{L'(1, \chi)}{L(1, \chi)}. \tag{10}$$

1.3 Ihara's Conjecture

Ihara made a conjecture on γ_m based on numerical observations for $m \leq 8000$, which we here only formulate in case $m = q$ is prime.

Conjecture 1 (Ihara's Conjecture [17]) Let $q \geq 3$ be a prime.

1) $\gamma_q > 0$ ('very likely');
2) For fixed $\epsilon > 0$ and q sufficiently large we have

$$\frac{1}{2} - \epsilon \leq \frac{\gamma_q}{\log q} \leq \frac{3}{2} + \epsilon.$$

The most extensive computations on γ_q to date were carried out by Ford et al. [7].

The largest value of $\gamma_q/\log q$ among $q \leq 30{,}000$ equals $1.626\ldots$ and occurs at $q = 19$. The smallest is $0.315\ldots$ and occurs at $q = 17{,}183$. It is a consequence of (5), (6) and Badzyan's result mentioned above that, under GRH, there exists a constant $c_2' > 0$ such that

$$-(1 - 1/\sqrt{5})q(\log q)/2 < \gamma_q < c_2' \log q. \tag{11}$$

Ihara [15] showed that $\gamma_q \leq (2 + o(1)) \log q$ assuming ERH (Conjecture 5 below). The lower bound in (11) turns out to be very weak. Ihara et al. [18] proved that for

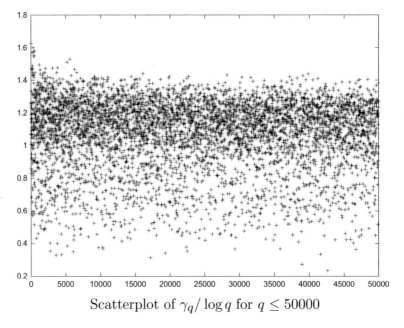

Scatterplot of $\gamma_q/\log q$ for $q \leq 50000$

Scatterplot of $\gamma_q/\log q$ for $q \leq 50000$

any $\epsilon > 0$ one has $|\gamma_q| = O_\epsilon(q^\epsilon)$ and, under GRH, $|\gamma_q| = O(\log^2 q)$. We will see in Sect. 3 that these bounds can be sharpened considerably.

1.4 Kummer's Conjecture

Let $h_1(q)$ be the ratio of the class number $h(q)$ of $\mathbb{Q}(\zeta_q)$ and the class number $h_2(q)$ of its maximal real subfield $\mathbb{Q}(\zeta_q + \zeta_q^{-1})$, that is, $h_1(q) = h(q)/h_2(q)$. Kummer proved that this is an integer. It is now called the first factor of the class number of $h(q)$.

In 1851 Kummer [19] published a review of the main results that he and others had discovered about cyclotomic fields. In this elegant report he made the following conjecture.

Conjecture 2 (Kummer's Conjecture [19]) Put

$$G(q) = \left(\frac{q}{4\pi^2}\right)^{\frac{q-1}{4}} \quad \text{and} \quad r(q) = \frac{h_1(q)}{G(q)}.$$

Then asymptotically $r(q)$ tends to 1.

In fact he claimed to have a proof that he would publish later together with further developments (but never did). Kummer himself laboriously computed $r(q)$ for $q < 100$. This was extended over time by many authors, more recently by Shokrollahi [33]. He showed that the largest value of $r(q)$ among $q \leq 10{,}000$ equals $1.556562\ldots$ and occurs at $q = 5231$. The smallest is $0.642429\ldots$ and occurs at $q = 3331$.

In 1949 Ankeny and Chowla [1, 2] made some progress by showing that

$$\log r(q) = o(\log q). \tag{12}$$

Siegel [34], who was unaware of the earlier work of Ankeny and Chowla, proved a weaker version of (12) and was one of the first to cast doubt on the truth of Kummer's conjecture. From (12) we infer that

$$\log h_1(q) \sim \frac{q}{4} \log q,$$

and thus that there are only finitely many primes q such that $\mathbb{Q}(\zeta_q)$ has class number one. This was made effective by Masley and Montgomery [22], who showed that $|\log r(q)| < 7 \log q$ for $q > 200$, which is strong enough to establish Kummer's conjecture that $h_1(q) = 1$ if and only if $q \leq 19$. This result is their key ingredient in determining all cyclotomic fields having class number 1. In proving their upper bound, Masley and Montgomery used zero-free regions of L-functions. This idea was refined by Puchta [32], with a further improvement by Debaene [5], to obtain an upper bound for $\log r(q)$ that depends on a Siegel zero, if it exists.

Not surprisingly h_1 is eventually monotonic; however, no beginning prime is yet known. In this direction Fung et al. [9, Theorem 1] showed that if E is an elliptic curve over $\mathbb{Q}$ for which the associated L-function has a zero of order at least 6 in $s = 1$, then it is possible to find an explicit prime q_0 for which $h_1(q_2) > h_1(q_1)$, whenever $q_2 > q_1 \geq q_0$. It is believed that one can take $q_0 = 19$.

Just like γ_q in (8), $h_1(q)$ is also related to special values of Dirichlet L-series. Hasse [14] showed that

$$r(q) = \frac{h_1(q)}{G(q)} = \prod_{\chi(-1)=-1} L(1, \chi), \tag{13}$$

where the product is over all the odd characters modulo q. It follows from this, (7) and (9) that

$$r(q) = \lim_{s \downarrow 1} \frac{\zeta_{\mathbb{Q}(\zeta_q)}(s)}{\zeta_{\mathbb{Q}(\zeta_q)^+}(s)}.$$

Indeed, using the definition of Euler-Kronecker constant we find the Taylor series expansion around $s = 1$

$$\frac{\zeta_{\mathbb{Q}(\zeta_q)}(s)}{\zeta_{\mathbb{Q}(\zeta_q)^+}(s)} = r(q)(1 + (\gamma_q - \gamma_q^+)(s-1) + O_q((s-1)^2)), \tag{14}$$

involving both of the main actors of this survey.[3]

1.5 Similarities Between the Two Conjectures

The remaining part of this survey will make clear that the quantities

$$\frac{\gamma_q}{\log q} \quad \text{and} \quad 1 - 2|\log r(q)| \tag{15}$$

have very similar analytical properties. Indeed, this analogy implies that the Euler-Kronecker analogue of the Kummer conjecture is that asymptotically

$$\gamma_q \sim \log q.$$

The numerical computations mentioned above suggest that both quantities in (15) are bounded, whereas if one believes in some standard conjectures in analytic number theory (delineated in the next section), they can be sporadically very negative. Various researchers in this area believe that it is the $\log \log \log$ devil that ruins both the Kummer and Ihara conjecture (see Sect. 5).

2 Preliminaries

2.1 Standard Conjectures Used

The results we are going to present depend on some standard conjectures on the prime distribution that we briefly recall in this section.

Let $\mathcal{A} = \{a_1, \ldots, a_s\}$ be a set consisting of s distinct natural numbers. We define

$$m(\mathcal{A}) = \sum_{k=1}^{s} \frac{1}{a_i}.$$

The set $\mathcal{A}$ is said to be admissible if there does not exist a prime p such that $p | \prod_{i=1}^{s}(a_i n + 1)$ for every $n \geq 1$. Note that if there is such a prime factor p, then $p \leq s + 1$. The sequence $\{a(i)\}_{i=1}^{\infty} = \{2, 6, 8, 12, 18, 20, 26, 30, 32, \ldots\}$ has

---

[3]I have not come across this formula in the literature.

the property that any finite subsequence is an admissible set. It is called "the greedy sequence of prime offsets" and is sequence A135311 in the Online Encyclopedia of Integer Sequences (OEIS).

Conjecture 3 (Hardy-Littlewood) Suppose $\mathcal{A} = \{a_1, \ldots, a_s\}$ is an admissible set. Then the number of primes $n \leq x$ such that the integers $a_1 n + 1, \ldots, a_s n + 1$ are all prime is $\gg_{\mathcal{A}} x \log^{-s-1} x$.

Actually, the full Hardy-Littlewood conjecture gives an asymptotic, rather than a lower bound. It is this full version that was used by Croot and Granville [4] to study how many primes $q \leq x$ satisfy $r(q) = \alpha + o(1)$, with $\alpha > 0$ and fixed.

As usual, we denote by $\pi(x; d, a)$ the number of primes $p \leq x$ satisfying $p \equiv a \pmod{q}$, $\pi(x)$ the prime counting function, $\mathrm{Li}(x)$ the logarithmic integral and μ the Möbius function.

Conjecture 4 (Elliott-Halberstam) For any $\epsilon > 0$ and $C > 0$ we have

$$\sum_{k < x^{1-\epsilon}} \max_{(l,k)=1} \max_{y \leq x} \left| \pi(y; k, l) - \frac{\mathrm{Li}(x)}{\varphi(k)} \right| \ll_{\epsilon, C} \frac{x}{(\log x)^C}.$$

Conjecture 5 (Extended Riemann Hypothesis) Every Dirichlet series $L(s, \chi)$ satisfies the Riemann Hypothesis.

Conjecture 6 (Generalized Riemann Hypothesis) Every Dedekind zeta function $\zeta_K(s)$ satisfies the Riemann Hypothesis.

In places where one uses GRH for a general number field, for a cyclotomic number field ERH suffices, as their Dedekind zeta function decomposes as a product of Dirichlet L-series, cf. (7).

For most results quoted below a weaker form of these conjectures suffices. For reasons of brevity we leave out the details and refer the reader to the original publications. Also for brevity we will refer to the above conjectures by the abbreviations HL, EH, ERH and GRH, respectively.

2.2 The Distribution of $m(\mathcal{A})$

Crucial for obtaining results on both the Kummer and the Ihara conjecture is an understanding of the distribution of $m(\mathcal{A})$ as $\mathcal{A}$ ranges over the admissible sets. We put $\mathcal{M} = \{m(\mathcal{A}) : \mathcal{A} \text{ is admissible}\}$ and let $\overline{\mathcal{M}}$ be the closure of $\mathcal{M}$, that is, the set of limit points of sequences of elements of $\mathcal{M}$ that do converge.

Granville showed that the following 1988 conjecture by Erdős is true.[4]

Theorem 1 (Granville [11]) *There is a sequence of admissible sets $A_1, A_2, \ldots$ such that $\lim_{j \to \infty} m(A_j) = \infty$.*

Corollary 1 *We have $\overline{\mathcal{M}} = [0, \infty]$.*

Proof Given any $x > 0$ and $\delta > 0$, there is an admissible set A with $m(A) > x$ consisting of integers all $> 1/\delta$. As any subset of an admissible set is also admissible, there is a subset A' of A with $|m(A') - x| < \delta$.

Another issue is that of finding admissible subsets $A \subseteq [1, x]$ having large $m(A)$. In this direction Granville proved the following result.

Proposition 1 (Granville [11])

1) *For any sufficiently large x there is an admissible set A, which is a subset of $[1, x]$, with $m(A) \geq (1 + o(1)) \log \log x$.*
2) *There exists a constant $c > 0$ such that if A is an admissible subset of $[1, x]$, then $m(A) \leq c \log \log x$.*

Granville believes one can take $c = 1 + \epsilon$, for any $\epsilon > 0$, provided that x is sufficiently large. If true, this would imply that part 1 is best possible.

3   The Constants γ_q: Results and Conjectures

On applying (4) and Lemma 1 we obtain

$$\gamma_q = -\frac{\log q}{q-1} - S(q) - \lim_{x \to \infty} \left(\log x - (q-1) \sum_{\substack{p \leq x \\ p \equiv 1 \ (\mathrm{mod}\ q)}} \frac{\log p}{p-1} \right), \qquad (16)$$

where

$$S(q) = (q-1) \sum_{\substack{p \neq q \\ \mathrm{ord}_p(q) \geq 2}} \frac{\log p}{p^{\mathrm{ord}_p(q)} - 1}.$$

By Lemma 1 the only rational primes splitting into prime ideals of prime norm are q and all the primes $p \equiv 1 \pmod{q}$. They are responsible for the first, respectively third term on the right-hand side of (16). The term $S(q)$ is the contribution of the

---

[4]The authors of [7], unaware of Granville's work and the fact that they were dealing with a(n) (ex-)conjecture of Erdős, gave a short different proof using a 1961 paper of…Erdős [6] himself. (The title of [6] has "G. Golomb" instead of the correct "S. Golomb".)

prime ideals lying above the remaining rational primes. Using estimates for linear forms in logarithms, it can be shown that $S(q) \leq 45$ and even that for any fixed $\epsilon > 0$ we have $S(q) < \epsilon$ for $(1 + o(1))\pi(x)$ primes $q \leq x$ [7, Theorem 3]. Since, as we will see, γ_q has normal order $\log q$, it follows that the first two terms in (16) are *error terms*.

The idea now is to approximate γ_q by choosing a suitable value for x in (16). In principle one wants to have x small, but the irregularities in the distribution of the primes do not allow us to take x too small. The Bombieri-Vinogradov theorem allows us to take $x = q^{2+\delta}$ for any $\delta > 0$ with the possible exception of a thin set of primes. Using the Brun-Titchmarsh inequality one can bring this down to $x = q^2$. Likewise, assuming EH one can go down to $x = q^{1+\delta}$. This approach leads to the following result.

Lemma 2 (Ford et al. [7]) *Given $r > 1$ write*

$$E_r(q) = \gamma_q - r \log q + q \sum_{\substack{p \leq q^r \\ p \equiv 1 \ (\mathrm{mod}\ q)}} \frac{\log p}{p - 1}. \tag{17}$$

1) *For all $C > 0$ we have $E_2(q) = O_C(\log \log q)$, with at most $O\left(\frac{\pi(x)}{(\log x)^C}\right)$ exceptions $q \leq x$.*
2) *Assuming EH, we have for fixed $\epsilon > 0$ and $C > 0$ that $E_{1+\epsilon}(q) = O_{C,\epsilon}(\log \log q)$, with at most $O\left(\frac{\pi(x)}{(\log x)^C}\right)$ exceptions $q \leq x$.*
3) *Assuming ERH, we have $E_2(q) = O(\log \log q)$.*

Before we consider how the large the prime sum in (17) with $0 < r \leq 2$ can be, we remark that it is usually small.

Proposition 2 (Ford et al. [7]) *Uniformly for $z \geq 2$, $\delta > 0$ and $0 < \epsilon \leq 1$, the number of primes $q \leq x$ for which*

$$q \sum_{\substack{p \leq q^{1+\epsilon} \\ p \equiv 1 \ (\mathrm{mod}\ q)}} \frac{\log p}{p - 1} \geq \delta \log q$$

is $O(\epsilon \pi(x)/\delta)$.

How small γ_q can be is determined by how large the prime sum in (17) can be.

Proposition 3 *There exists an absolute constant $c > 0$ such that on a set of primes of natural density 1 we have*

$$-c \log \log q < \frac{\gamma_q}{\log q} < (2 + \epsilon) \log q,$$

with $\epsilon > 0$ arbitrary and fixed.
    Under ERH these estimates hold for all primes q large enough.

Proof On writing the primes $p \equiv 1 \pmod{q}$ that satisfy $p \leq q^2$ as $a_1 q + 1, \ldots, a_s q + 1$, and noting that $\mathcal{A} := \{a_1, \ldots, a_s\}$ is an admissible set, we obtain by Proposition 1 that

$$q \sum_{\substack{p \leq q^2 \\ p \equiv 1 \pmod{q}}} \frac{\log p}{p-1} < 2m(\mathcal{A}) \log q \ll \log q \log \log q.$$

Now the unconditional statement follows on invoking part 1 of Lemma 2.

Under ERH the upper bound is due to Ihara [15] and the lower bound to Badzyan[5] [3].

In the next section we will see that for $r = 2$ the prime sum in (17) can be quite large if we assume HL.

3.1 Assuming HL

Armed with HL and Lemma 2, it is easy to give a conditional disproof of part 1 of Ihara's conjecture.

Theorem 2 *Suppose that HL is true and that $\mathcal{A}$ is an admissible set. Then one has*

$$\gamma_q < (2 - m(\mathcal{A}) + o(1)) \log q$$

for $\gg x \log^{-\#\mathcal{A}-1} x$ primes $q \leq x$.

Proof Let $a_1, \ldots, a_s$ be the elements of $\mathcal{A}$. By HL there are infinitely many primes q such that infinitely often $a_1 q + 1, \ldots, a_s q + 1$ are all prime and in addition $a_s q + 1 \leq q^2$. Then

$$q \sum_{\substack{p \leq q^2 \\ p \equiv 1 \pmod{q}}} \frac{\log p}{p-1} > q \sum_{i=1}^{s} \frac{\log q}{a_i q} = m(\mathcal{A}) \log q.$$

The proof now easily follows from the part 1 of Lemma 2 with any $C > s$.

A computer calculation gives that $\mathcal{A} = \{a(1), \ldots, a(2088)\}$ satisfies $m(\mathcal{A}) > 2$, where $a(1), a(2), \ldots$ is the sequence of integers introduced in Sect. 2.1. We thus obtain the following corollary of Theorem 2.

Corollary 2 *Assume HL. Then part 1 of Ihara's conjecture is false for infinitely many primes q.*

---

[5]He assumes GRH. The reproof given in [7, p. 1470] shows that ERH is sufficient.

Unconditionally we only have the following result.

Theorem 3 (Ford et al. [7]) *We have $\gamma_{964477901} = -0.1823\ldots$, and so part 1 of Ihara's conjecture is false for at least one prime q.*

This looks perhaps easy, but was made possible only by a new, fast algorithm developed by the authors of [7] (it requires computation of $L(1, \chi)$ for all characters modulo q). The prime $q = 964{,}477{,}901$ has the property that $aq + 1$ is prime for $a \in \{2, 6, 8, 12, 18, 20, 26, 30, 36, 56, \ldots\}$. It is easy to approximate the above value of γ_q by taking a large x in formula (16). The authors of [7] believe that if there is a further q with $\gamma_q < 0$, then its computation will be hopelessly infeasible.

Since by Theorem 1 one can find admissible $\mathcal{A}$ with $m(\mathcal{A})$ arbitrarily large, we obtain the following result from Theorem 2.

Theorem 4 (Ford et al. [7]) *Assume HL. Then*

$$\lim_{q \to \infty} \inf \frac{\gamma_q}{\log q} = -\infty.$$

Thus, conditionally, γ_q can be very negative. This happens not frequently since Mourtada and Murty [27] showed unconditionally that the set of primes $q \leq x$ such that $\gamma_q \leq -11 \log q$ is of size $o(\pi(x))$. In Sect. 5 we speculate how negative γ_q as a function of q can be.

3.2 Assuming EH (and HL)

The prime sum in (17) cannot be too small by Proposition 2, and on invoking the part 2 of Lemma 2 we obtain the following result.

Theorem 5 (Ford et al. [7]) *Assume EH. Let $\epsilon > 0$ be arbitrary. For a density 1 sequence of primes q we have*

$$1 - \epsilon < \frac{\gamma_q}{\log q} < 1 + \epsilon.$$

This describes the situation for the bulk of the primes. However, if one assumes in addition HL, one can say something about the irregular behaviour.

Theorem 6 *Suppose that both EH and HL are true. If $\mathcal{A}$ is an admissible set, then one has*

$$\gamma_q = (1 - m(\mathcal{A}) + o(1)) \log q$$

for $\gg_{\mathcal{A}} x \log^{-\#\mathcal{A}-1} x$ primes $q \leq x$.

Proof (Sketch of Proof) By reasoning as in the proof of Theorem 2, we obtain $\gamma_q \leq (1 - m(\mathcal{A}) + o(1)) \log q$. The reverse inequality is obtained on using sieve methods to find enough primes $q \leq x$ with $qa + 1$ prime for $a \in \mathcal{A}$ and not prime for $a \notin \mathcal{A}$ and $a \leq q^{\epsilon}$, see [7, p. 1465] for details.

Now using that $\overline{\mathcal{M}} = [0, \infty]$ (Corollary 1), we obtain the following result.

Theorem 7 (Ford et al. [7]) *Assume EH and HL. Then the set*

$$\Theta := \left\{ \frac{\gamma_q}{\log q} : q \text{ prime} \right\}$$

is dense in $(-\infty, 1]$.

We propose the following conjecture.

Conjecture 7 Let $\mathcal{A}$ be any admissible set. If EH and HL are both true, then $1 - m(\mathcal{A})$ is a limit point of the set Θ.

3.3 Cyclotomic Euler-Kronecker Constants on Average

Murty [28] proved unconditionally that

$$\sum_{Q/2 < q \leq Q} |\gamma_q| \ll (\pi(Q) - \pi(Q/2)) \log Q.$$

Fouvry [8] showed that uniformly for $M \geq 3$ one has the equality

$$\frac{1}{M} \sum_{M/2 < m \leq M} |\gamma_m| = \log M + O(\log \log M),$$

if one ranges over the integers m, rather than the primes q.

4 The Kummer Conjecture: (Conditional) Results

The orthogonality property of characters gives us

$$\sum_{\chi(-1)=-1} \log L(s, \chi) = \frac{q-1}{2} \sum_{p^m \equiv \pm 1 \pmod{q}} \pm \frac{1}{mp^{ms}},$$

where the latter notation is shorthand for

$$\sum_{\substack{p^m \equiv 1 \ (\text{mod } q)}} \frac{1}{mp^{ms}} - \sum_{\substack{p^m \equiv -1 \ (\text{mod } q)}} \frac{1}{mp^{ms}}.$$

From Hasse's formula (13) we have that

$$\log r(q) = \frac{q-1}{2} \lim_{x \to \infty} \left(\sum_{m \geq 1} \frac{1}{m} \sum_{\substack{p^m \leq x \\ p^m \equiv \pm 1 \ (\text{mod } q)}} \pm \frac{1}{p^m} \right). \tag{18}$$

We denote the limit by f_q. Note that Kummer's conjecture is equivalent with $f_q = o(1/q)$. Formula (18) should be compared to formula (16). As in that formula, one tries to choose x as small as possible so that the resulting error is still reasonable. In doing so, also here Bombieri-Vinogradov theorem and Brun-Titchmarsh inequality come into play. The main contribution to f_q comes from the term with $m = 1$. Taking all this into account, Granville [11] showed that if Kummer's conjecture is true, then for every $\delta > 0$ we must have

$$\sum_{\substack{p \leq q^{1+\delta} \\ p \equiv \pm 1 \ (\text{mod } q)}} \pm \frac{1}{p} = o\left(\frac{1}{q}\right),$$

for all but at most $\ll x / \log^3 x$ primes $q \leq x$.

Using this approach Granville showed that

$$1/c \leq r(q) \leq c$$

for a positive proportion $\rho(c)$ of primes $p \leq x$, where $\rho(c) \to 1$ as $c \to \infty$. Murty and Petridis [29] improved this as follows.

Theorem 8 *There exists a positive constant c such that for a sequence of primes with natural density 1 we have*

$$c^{-1} \leq r(q) \leq c.$$

If EH is true, then we can take $c = 1 + \epsilon$ for any fixed $\epsilon > 0$.

Thus Ram Murty and Yiannis Petridis showed that a weaker version of Kummer's conjecture holds true. Yet, if both EH and HL are true, Kummer's conjecture itself is false and, moreover, we have the following much stronger result.

Theorem 9 (Granville [11]) *Put*

$$\Omega = \{r(q) : q \text{ is prime}\}.$$

Assume both HL and EH. Then the sequence Ω has $[0, \infty]$ as set of limit points.

This result follows from Corollary 1 and the following.

Theorem 10 *If EH and HL are both true, then, for any admissible set $\mathcal{A}$, the numbers $e^{m(\mathcal{A})/2}$ and $e^{-m(\mathcal{A})/2}$ are both limit points of Ω.*

5 The Log Log Log Devil Unleashed

Regarding the extremal behaviour of $r(q)$ and $\gamma_q / \log q$, we enter the realm of speculation, following Granville [11, Section 9].

Speculation 1 (Granville [11]) *For all primes q, we have*

$$(-1 + o(1)) \log \log \log q \leq 2 \log r(q) \leq (1 + o(1)) \log \log \log q. \tag{19}$$

These bounds are best possible in the sense that there exist two infinite sequences of primes for which the lower, respectively upper bound are attained.

The same line of thought for γ_q gives rise to the following speculation.

Speculation 2 *For all primes q, we have*

$$\frac{\gamma_q}{\log q} \geq (-1 + o(1)) \log \log \log q. \tag{20}$$

The bound is best possible in the sense that there exists an infinite sequence of primes for which the bound is attained.

We will now sketch the motivation for these two speculations and do this in parallel, to bring out the analogy in the reasoning more clearly. The speculations require the assumption that primes are both more regularly and more irregularly distributed than can be currently established.

For convenience let us write $L_2 = \log \log q$ and $L_3 = \log \log \log q$. We assume that there exists an absolute constant $A > 0$ for which we can take $x = q(\log q)^A$ in (16), such that the estimate

$$\gamma_q = \log q - q \sum_{\substack{p \leq q(\log q)^A \\ p \equiv 1 \pmod{q}}} \frac{\log p}{p - 1} + E(q), \tag{21}$$

with $E(q) = o(L_3 \log q)$ holds true. Now note that

$$\sum_{\substack{p \leq q(\log q)^A \\ p \equiv 1 \pmod{q}}} \frac{\log p}{p - 1} = \sum_{\substack{p \leq q(\log q)^A \\ p \equiv 1 \pmod{q}}} \frac{\log q}{p} \left(1 + O_A\left(\frac{L_2}{\log q}\right)\right), \tag{22}$$

where we used that the log p appearing on the left-hand side of (22) satisfies $\log p = \log q + O_A(L_2)$. Combining (22) and (21) then yields

$$\frac{\gamma_q}{\log q} = 1 - \sum_{\substack{2q-1 \leq p \leq q(\log q)^3 \\ p \equiv 1 \pmod q}} \frac{q}{p}\left(1 + O_A\left(\frac{L_2}{\log q}\right)\right) + \frac{E(q)}{\log q}. \tag{23}$$

Granville [11, p. 335] makes some speculations about the distribution of prime numbers that would ensure that one can go down to $x = q(\log q)^3$ in the Kummer problem and lead to[6]

$$\log r(q) = \frac{q-1}{2} \sum_{\substack{p \leq q(\log q)^3 \\ p \equiv \pm 1 \pmod q}} \pm \frac{1}{p} + O\left(\frac{1}{\sqrt{\log q}}\right). \tag{24}$$

By the Brun-Titchmarsh theorem there exists a constant $c > 0$ such that for all $x \geq 2q - 1$ we have

$$\max\{\pi(x; q, -1), \pi(x; q, 1)\} \leq c \frac{x}{(q-1)\log(x/q)}.$$

Using this it is easy to deduce that

$$\sum_{\substack{2q+1 \leq p \leq q(\log q)^A \\ p \equiv 1 \pmod q}} \frac{1}{p} \leq \frac{c}{q}(L_3 + O_A(1)). \tag{25}$$

Combining this estimate with (23) gives

$$\frac{\gamma_q}{\log q} \geq 1 - cL_3 + O_A(1) + \frac{E(q)}{\log q}. \tag{26}$$

Similarly, combining (25) with $A = 3$ and (24) yields

$$\log r(q) \leq cL_3/2 + O(1). \tag{27}$$

It follows from (24) that

$$\log r(q) \geq -\frac{q-1}{2} \sum_{\substack{2q-1 \leq p \leq q(\log q)^3 \\ p \equiv -1 \pmod q}} \frac{1}{p} + O\left(\frac{1}{\sqrt{\log q}}\right),$$

and a similar argument as before now yields

$$\log r(q) \geq -cL_3/2 + O(1). \tag{28}$$

---

[6] Having the larger error term $o(L_3)$ would also suffice for our purposes.

Montgomery and Vaughan [24] have shown that we may take $c = 2$ and it is conjectured that one may take $c = 1 + o(1)$. If this is so, then combining (27) with (28) yields (19). Likewise, (26) gives rise to the lower bound (20).

The final step is to argue why the bounds (19) and (20) are best possible. We will only do so for the easier case of the bound (20). We assume the Hardy-Littlewood conjecture in a stronger form, namely in its original asymptotic form. Then one can argue that for any admissible set $\mathcal{A}$ with elements $\leq z$, we only find enough primes q for which $p = qa + 1$ is prime for all $a \in \mathcal{A}$ if $q > z^{10z}$ and z is large enough. This in combination with the part 1 of Proposition 1 then suggests that there are infinitely many primes q for which

$$\sum_{\substack{2q+1 \leq p \leq q(\log q)^A \\ p \equiv 1 \pmod{q}}} \frac{1}{p} \geq (1 - \epsilon)\frac{L_3}{q}. \tag{29}$$

This estimate, together with (23) and the already obtained lower bound (20), then finishes our argumentation.

These two speculations taken together imply the following weaker one.

Speculation 3 *There exists a function $g(q)$ such that*

$$\lim_{q \to \infty} \inf \frac{\gamma_q}{g(q) \log q} = 2 \lim_{q \to \infty} \inf \frac{\log r(q)}{g(q)} < 0.$$

In case $g(q)$ is not the log log log devil from the title, it is certainly a close cousin!

Comparison of Conjecture 7 and Theorem 10 suggests that $\gamma_q / \log q$ and $1 - 2|\log r(q)|$ behave similarly, which is consistent with the three speculations presented in this section.

6 Prospect

6.1 Polymath

Recent progress on gaps between primes allows one to meet the challenge below for some $C > 0$. Indeed, according to Maynard [23], recent results allow one to take $C = 1/246$.

Challenge 1 *Find a set $\mathcal{A} = \{a_1, \ldots, a_s\}$ such that provably for some $B > 0$ there are $\gg x / \log^B x$ primes $q \leq x$ such that $a_1 q + 1, \ldots, a_s q + 1$ are all prime and, in addition,*

$$\sum_{i=1}^{s} \frac{1}{a_i} \geq C,$$

with C as large as possible.

Conjecturally C can be taken arbitrarily large, cf. Theorem 1.

Proposition 4 *If one meets the challenge for any $C > 2$, then there are $\gg$ $x / \log^B x$ primes $q \leq x$ for which part 1 of Ihara's conjecture is false and, moreover,* $\gamma_q < (2 - C) \log q$.

Proof Similar to that of Theorem 2.

6.2 Kummer for Arbitrary Cyclotomic Fields

It is not difficult to formulate a generalized Kummer conjecture, where instead of the primes we range over the integers. Goldstein [10] established that, as r tends to infinity and q is fixed, we have

$$\log h_1(q^r) \sim \frac{r}{4}\left(1 - \frac{1}{q}\right) q^r \log q.$$

Myers [30] obtained some results along the lines of Murty and Petridis [29]. Fouvry [8] determined the average order of $|\gamma_m|$ (see Sect. 3.3). Quite likely further results can be obtained, e.g., it is perhaps possible to find explicit composite integers m for which $\gamma_m < 0$.

6.3 Strengthening the Analogy (Moree and Saad Eddin [26])

Comparison of (8) and (13) suggests that one can expect an even closer analogy between $r(q)$ and the difference

$$\gamma_q - \gamma_q^+ = \sum_{\chi(-1)=-1} \frac{L'(1, \chi)}{L(1, \chi)}, \tag{30}$$

which results on subtracting (10) from (8). In particular, it is to be expected that $\gamma_q - \gamma_q^+$ will display, like $\log r(q)$, a more symmetric behaviour around the origin than γ_q does. Also recall that $r(q)$ and $\gamma_q - \gamma_q^+$ both appear in the Taylor series (14).

Acknowledgements I would like to thank James Maynard for pointing out that one can take $C = 1/246$ in Challenge 1. Furthermore, I am grateful to Alexandru Ciolan, Sumaia Saad Eddin and Alisa Sedunova for proofreading and help with editing an earlier version. Ignazio Longhi and the referee kindly pointed out some disturbing typos.

The similarity between Kummer's and Ihara's conjectures was pointed out by Andrew Granville after a talk given by Kevin Ford on [7]. At that point the authors of [7] had independently obtained Theorem 1, but not Granville's Proposition 1, the latter being precisely the result used by Granville to unleash the log log log devil. Once at the loose, it created havoc also among the Euler-Kronecker constants.

References

1. N.C. Ankeny, S. Chowla, The class number of the cyclotomic field. Proc. Natl. Acad. Sci. U. S. A. **35**, 529–532 (1949)
2. N.C. Ankeny, S. Chowla, The class number of the cyclotomic field. Can. J. Math. **3**, 486–494 (1951)
3. A.I. Badzyan, The Euler–Kronecker constant. Mat. Zametki **87**, 45–57 (2010). English Translation in Math. Notes **87**, 31–42 (2010)
4. E.S. Croot III, A. Granville, Unit fractions and the class number of a cyclotomic field. J. Lond. Math. Soc. (2) **66**, 579–591 (2002)
5. K. Debaene, The first factor of the class number of the p-th cyclotomic field. Arch. Math. (Basel) **102**, 237–244 (2014)
6. P. Erdős, On a problem of S. Golomb. J. Aust. Math. Soc. **2**, 1–8 (1961/1962)
7. K. Ford, F. Luca, P. Moree, Values of the Euler ϕ-function not divisible by a given odd prime, and the distribution of Euler-Kronecker constants for cyclotomic fields. Math. Comput. **83**, 1447–1476 (2014)
8. É. Fouvry, Sum of Euler-Kronecker constants over consecutive cyclotomic fields. J. Number Theory **133**, 1346–1361 (2013)
9. G. Fung, A. Granville, H.C. Williams, Computation of the first factor of the class number of cyclotomic fields. J. Number Theory **42**, 297–312 (1992)
10. L.J. Goldstein, On the class numbers of cyclotomic fields. J. Number Theory **5**, 58–63 (1973)
11. A. Granville, On the size of the first factor of the class number of a cyclotomic field. Invent. Math. **100**, 321–338 (1990)
12. Y. Hashimoto, Euler constants of Euler products. J. Ramanujan Math. Soc. **19**, 1–14 (2004)
13. Y. Hashimoto, Y. Iijima, N. Kurokawa, M. Wakayama, Euler's constants for the Selberg and the Dedekind zeta functions. Bull. Belg. Math. Soc. Simon Stevin **11**, 493–516 (2004)
14. H. Hasse, *Über die Klassenzahl Abelscher Zahlkörper*. Mathematische Lehr- bücher und Monographien, Band I (Akademie-Verlag, Berlin, 1952)
15. Y. Ihara, On the Euler-Kronecker constants of global fields and primes with small norms, in *Algebraic Geometry and Number Theory: In Honor of Vladimir Drinfeld's 50th Birthday*, ed. by V. Ginzburg. Progress in Mathematics, vol. 850 (Birkhäuser, Boston, Cambridge, MA, 2006), pp. 407–451
16. Y. Ihara, On "M-functions" closely related to the distribution of L'/L-values. Publ. Res. Inst. Math. Sci. **44**, 893–954 (2008)
17. Y. Ihara, The Euler-Kronecker invariants in various families of global fields, in *Proceedings of Arithmetic Geometry and Coding Theory 10 (AGCT 2005), Séminaires et Congrès*, ed. by F. Rodier et al., vol. 21 (2009), pp. 79–102
18. Y. Ihara, V.K Murty, M. Shimura, On the logarithmic derivatives of Dirichlet L-functions at $s = 1$. Acta Arith. **137**, 253–276 (2009)
19. E.E. Kummer, Mémoire sur la théorie des nombres complexes composées de racines de l'unité et des nombres entiers. J. Math. Pures Appl. **16**, 377–498 (1851). Collected Works, Vol. I., pp. 363–484
20. J.C. Lagarias, Euler's constant: Euler's work and modern developments. Bull. Am. Math. Soc. (N.S.) **50**, 527–628 (2013)
21. S. Lang, *Cyclotomic Fields I and II*, Combined 2nd edn. Graduate Texts in Mathematics, vol. 121 (Springer, New York, 1990)
22. J.M. Masley, H.L. Montgomery, Cyclotomic fields with unique factorization. J. Reine Angew. Math. **286/287**, 248–256 (1976)
23. J. Maynard, E-mail to author, 04/14/2014
24. H.L. Montgomery, R.C. Vaughan, The large sieve. Mathematika **20**, 119–134 (1973)
25. P. Moree, Counting numbers in multiplicative sets: Landau versus Ramanujan. Math. Newsl. **21**(3), 73–81 (2011). arXiv:1110.0708

26. P. Moree, S. Saad Eddin, A. Sedunova, Euler-Kronecker constants for maximal real cyclotomic fields and Kummer's conjecture (in preparation)
27. M. Mourtada, V.K. Murty, On the Euler Kronecker constant of a cyclotomic field, II, in *SCHOLAR–A Scientific Celebration Highlighting Open Lines of Arithmetic Research: Centre de Recherches Mathematiques Proceedings*. Contemporary Mathematics, vol. 655 (American Mathematical Society, Providence, RI, 2015), pp. 143–151
28. V.K. Murty, The Euler-Kronecker constant of a number field. Ann. Sci. Math. Que. **35**, 239–247 (2011)
29. M.R. Murty, Y.N. Petridis, On Kummer's conjecture. J. Number Theory **90**, 294–303 (2001)
30. M.J.R. Myers, A generalised Kummer's conjecture. Glasg. Math. J. **52**, 453–472 (2010)
31. W. Narkiewicz, *Elementary and Analytic Theory of Algebraic Numbers*, 2nd edn. (Springer, Berlin; PWN—Polish Scientific Publishers, Warsaw, 1990)
32. J.-C. Puchta, On the class number of p-th cyclotomic field, Arch. Math. (Basel) **74**, 266–268 (2000)
33. M.A. Shokrollahi, Relative class number of imaginary abelian fields of prime conductor below 10000, Math. Comput. **68**, 1717–1728 (1999)
34. C.L. Siegel, Zu zwei Bemerkungen Kummers. Nachr. Akad. Wiss. Göttingen Math.-Phys. Kl. II **6**, 51–57 (1964). Collected Works III, 438–442
35. M.A. Tsfasman, Asymptotic behaviour of the Euler-Kronecker constant, in *Algebraic Geometry and Number Theory: In Honor of Vladimir Drinfeld's 50th Birthday*, ed. by V. Ginzburg. Progress in Mathematics, vol. 850 (Birkhäuser, Boston, Cambridge, MA, 2006), pp. 453–458
36. L.C. Washington, *Introduction to Cyclotomic Fields*. Graduate Texts in Mathematics, vol. 83 (Springer, New York, 1982)

Maier's Matrix Method and Irregularities in the Distribution of Prime Numbers

Andrei Raigorodskii and Michael Th. Rassias

Abstract This paper is devoted to irregularities in the distribution of prime numbers. We describe the development of this theory and the relation to Maier's matrix method.

1 Introduction

In the paper [13] Helmut Maier for the first time applied the now famous Maier's matrix method to obtain important results on chains of large gaps between consecutive primes. In a subsequent paper [14] he used this method to obtain surprising irregularity results on the distribution of prime numbers in short intervals. These irregularity results were later extended to other situations in various papers.

In this paper we describe the basic features of Maier's matrix method as well as the irregularity results obtained from it. There have been other applications of the matrix method, like strings of consecutive primes in the same residue class, which we do not cover in detail. We refer the interested reader to the papers [15] and [21]. Pictures of the matrices used may also be found in the survey article [9]. Many topics discussed in our paper may also be found there.

A. Raigorodskii
Moscow Institute of Physics and Technology, Dolgoprudny, Russia

Moscow State University, Moscow, Russia

Buryat State University, Ulan-Ude, Russia
e-mail: raigorodsky@yandex-team.ru

M. Th. Rassias (✉)
Institute of Mathematics, University of Zürich, Zürich, Switzerland

Moscow Institute of Physics and Technology, Dolgoprudny, Russia

Institute for Advanced Study, Program in Interdisciplinary Studies, Princeton, NJ, USA
e-mail: michail.rassias@math.uzh.ch; rassias@ias.edu

© Springer International Publishing AG, part of Springer Nature 2018
J. Pintz, M. Th. Rassias (eds.), *Irregularities in the Distribution of Prime Numbers*,
https://doi.org/10.1007/978-3-319-92777-0_9

2   The Maier Matrix Method

A key ingredient in many applications of Maier's matrix method, the sequence of sieving steps, has first been developed in papers of Erdős [4] and Rankin [19] on large gaps between consecutive primes. There are five sieving steps in this paper.

The sifted set is contained in a single interval $(0, P(x)]$ of consecutive integers, where

$$P(x) = \prod_{p<x} p \ .$$

Let

$$y = C\, x \left(\log x \ \frac{\log_3 x}{\log_2^2 x} \right),$$

where C is a constant. By the sieving a subinterval $(s, s + y]$ of $(0, P(x)]$ is constructed containing no integers relatively prime to $P(x)$. Each sieving step consists in the introduction of congruence conditions

$$s \equiv a_p \mod p \ , \quad \text{for all } p \in P_i \ , \tag{2.1}$$

where P_i is a subset of

$$\{p \le x, \ p \text{ prime}\} \text{ for } 1 \le i \le 5 \ .$$

The number s will only be determined uniquely mod $P(x)$ after all five sieving steps have been completed.

Let the set of s satisfying (2.1) for $1 \le i \le j$ be denoted by T_j. Then $s \in T_j$ implies that $s + g$ is composite for $g \in (0, y]$ except for a set S_j of *survivors*. In each sieving step the set of survivors is reduced, i.e. $S_{j+1} \subset S_j$.

In the papers of Erdős [4] and Rankin [19] we have $S_5 = \emptyset$. Thus for the value of s, uniquely determined by

$$s \equiv a_p \mod p \ \text{ for all } p \le x \text{ and } 0 < s \le P(x) \ ,$$

the interval $(s, s + y]$ contains no prime numbers, which gives the results of [4] and [19].

In the first appearance of the matrix method in the paper [13], the first step consists in the construction of a base row $B = (s, s + y]$. This base row is obtained almost in the same manner as the subinterval $(s, s + y]$ of $(0, P(x)]$ in the papers of Erdős and Rankin described previously. The only difference is that the last sieving step is not carried out. Thus, the base row B contains a non-empty set of survivors of the previous sieving steps, a set of integers coprime to $P(x)$ with the appropriate density.

The rows of the matrix $\mathcal{M}$ now are translates $B + rP(x)$ of the base row B. We have

$$\mathcal{M} = (a_{r,u}) \text{ with } a_{r,u} = u + rP(x), \ P(x)^{D-1} < r \leq 2P(x)^{D-1}, \ u \in (s, s+y].$$

The columns of this matrix

$$\mathcal{C}(u) = \left\{ u + rP(x) \ : \ P(x)^{D-1} < r \leq 2P(x)^{D-1} \right\}$$

are arithmetic progressions. Only the *admissible columns* $\mathcal{C}(u)$ with $(u, P(x)) = 1$ can contain primes. The number of primes in these admissible columns can be studied by the application of appropriate prime-number theorems for arithmetic progressions. By estimating the total number of primes in the matrix $\mathcal{M}$, it can be shown that there are many rows containing at least the appropriate number of primes.

The concentration of the primes on few rows and also the occurrence of small gaps between the primes in many rows is ruled out by the application of sieve results on upper bounds for the number of generalized twin primes.

3 Irregularity Results for Primes in Short Intervals

We start with some definitions, measuring the irregularities in the distribution of primes. We follow [5].

Definition 3.1 For $x \geq 2$, set

$$\pi(x) = \sum_{p \leq x} 1 = (1 + \Delta_1(x)) \int_0^x \frac{dt}{\log t}$$

$$\theta(x) = \sum_{p \leq x} \log p = (1 + \Delta_2(x))x$$

$$\psi(x) = \sum_{p^m \leq x} \log p = (1 + \Delta_3(x))x \ .$$

To have a short notation for irregularities in short intervals we introduce the expressions $\Delta_i(x, y)$ by:

Definition 3.2 For $x \geq 2$, $y \geq 1$, we set

$$\pi(x + y) - \pi(x) = (1 + \Delta_1(x, y)) \int_x^{x+y} \frac{dt}{\log t}$$

$$\theta(x + y) - \theta(x) = (1 + \Delta_2(x, y))y$$

$$\psi(x + y) - \psi(x) = (1 + \Delta_3(x, y))y .$$

Before discussing irregularities and unexpected behavior of the quantities in Definition 3.2 one should first have some thoughts, what the expected behavior is. Here a basic role is played by Cramér's model [2]:

Define the independent random variables X_n ($n \geq 3$) such that

$$X_n = \begin{cases} 1 \text{ with probability } 1/\log n \\ 0 \text{ with probability } 1 - 1/\log n \end{cases}$$

It may be expected that a property that is true for the sequence $(X_n)_{n \geq 3}$ with probability 1 will hold also for the sequence of primes.

In analogy to the functions $\Delta_i(x, y)$ let us define for $x \geq 3$ and $y \geq 1$ random variables $D_i(x, y)$ ($i = 1, 2$) by setting

$$\sum_{x < n \leq x+y} X_n = (1 + D_1(x, y)) \int_x^{x+y} \frac{dt}{\log t}$$

$$\sum_{x < n \leq x+y} X_n \log n = (1 + D_2(x, y))y .$$

Then the following result holds.

Theorem 3.3 (Cramér's Model) *With probability 1 we have for $i = 1, 2$*

$$D_i(x, y) \ll \frac{\log x}{\sqrt{y}} \tag{3.1}$$

uniformly in the range

$$x \geq 3, \ 10(\log x)^2 \leq y \leq x .$$

The relation (3.1) suggests that

$$\Delta_1\left(x, (\log x)^\lambda\right) = o(1), \ (x \to \infty), \text{ if } \lambda > 2 . \tag{3.2}$$

In the paper [14] Maier proves a result contradicting the expected result (3.2):
Let $\lambda \in (1, \infty)$ be fixed. Then we have:

$$\Delta_1(x, (\log x)^\lambda) = \Omega_\pm(1) , \tag{3.3}$$

(from which analogous results for Δ_2, Δ_3 follow).

In the sequel we sketch the proof of (3.3). To prepare an appropriate base row for the matrix $\mathcal{M}$ one uses oscillation results on sifting functions. The following function is well-known.

Definition 3.4 For $w \geq z \geq 2$ we set

$$\Phi(w, z) = \#\{n \leq z \ : \ p \mid n \Rightarrow p \geq w\}.$$

This function is a special case of the function $\Phi_k(w, z)$.

Definition 3.5 For $k \in \mathbb{N}$, $w \geq z \geq 2$, we set

$$\Phi_k(w, z) = \#\{n \leq y \ : \ p \mid n, \ p \nmid k \Rightarrow p \geq z\}.$$

The function $r_k(w, z)$, measuring the oscillation of $\Phi_k(w, z)$ is defined by

$$\Phi_k(w, z) = w \prod_{\substack{p < z \\ p \nmid k}} \left(1 - \frac{1}{p}\right)(1 + r_k(y, z)). \tag{3.4}$$

We have $\Phi_1(w, z) = \Phi(w, z)$ and also set

$$r(y, z) = r_1(y, z).$$

For $k = 1$, $y = z^\lambda$, we have:

$$\lim_{z \to \infty} r_k(y, z) = e^\gamma \omega(\lambda) - 1,$$

where $\omega(u)$ is defined by

$$\begin{cases} \omega(u) = u^{-1}, \ 1 \leq u \leq 2 \\ \frac{d}{du}(u\,\omega(u)) = \omega(u - 1), \ u \geq 2, \end{cases} \tag{3.5}$$

and where the right-hand derivative has to be taken at $u = 2$.

The needed oscillation result is a special case of a general result of Iwaniec [12] on the sieve of Eratosthenes:

The function

$$\omega(u) - e^{-\gamma}$$

changes sign in any interval $[u - 1, u)$, $u \geq 2$.

We choose x, such that there are no *bad* Siegel zeros mod $P(x)$. A theorem of Gallagher [8] then guarantees that for R of order $P(x)^D$, D sufficiently large, the finite arithmetic progressions

$$((R + i)P + j)_{1 \leq i \leq R - 1}$$

contain the expected number of prime numbers. The matrix method is applied with
the matrix

$$\mathcal{M} = ((R+i)P + j)_{\substack{1 \le i \le R-1 \\ 1 \le j \le y}}, \text{ where } y = (\log(RP))^\lambda .$$

Thus, $\mathcal{M}$ consists of translates of the base row B. By (3.4) and (3.5), $\mathcal{M}$ contains

$$y \prod_{p<x} \left(1 - \frac{1}{p}\right) (e^\gamma \omega(\lambda) - 1)(1 + o(1)) \ (x \to \infty)$$

admissible columns.

In this irregularity result the oscillations of the sifting functions are fairly large.
The existence of smaller oscillations can be established for much longer intervals.
This has been carried out in joint work between A. Hildebrand and H. Maier [11].
Here the intervals $(x, x + y]$ containing an unusual number of primes may have
length as large as

$$y = \exp\left((\log x)^{1/3-\epsilon}\right), \ (\epsilon > 0 \text{ fixed}) .$$

The following result is proved (Theorem 1 of [11]):

Let $i \in \{1, 2, 3\}$ and $0 < \epsilon < 1/2$ be fixed. Then, for all sufficiently large values
of x and

$$2 \le y \le \exp\left(A(\log x)^{1/3}\right)$$

there exist numbers $x_\pm$ satisfying

$$x \le x_\pm \le 2x$$

such that

$$\pm \Delta_i(x_\pm, y) \ge c_0 y^{-(1+\epsilon)\delta(x,y)} ,$$

where

$$\delta(x, y) = \frac{\log(\log y / \log \log x)}{\log(\log x / \log y)} ,$$

A is a positive absolute constant and $c_0 = c_0(\epsilon)$ a positive constant depending at
most on ϵ.

This result also contains the result (3.3). For the statement of the oscillation result
we give the following definition, which we borrow from [11].

Definition 3.6 Given a positive integer q, we define for $x \geq 0$ and $y \geq 1$, a function $r(x, y; q)$ by

$$\sum_{\substack{x < n \leq x+y \\ (n,q)=1}} 1 = (1 + r(x, y; q)) \frac{\phi(q)}{q} y \,,$$

and set

$$r^*(y; q) = \max_{x \geq 0} r(x, y; q) \,,$$

$$r_*(y; q) = \min_{x \geq 0} r(x, y; q) \,.$$

For a prime p_0, we set

$$P(z; p_0) = \prod_{p \neq p_0} p \,.$$

The oscillation result now is the following (Proposition 3 of [11]):

Let $0 < \epsilon < 1/2$ be given. Then, for sufficiently large z, $2 \leq y \leq \exp \sqrt{z}$, and every prime p_0, we have

$$\left. \begin{array}{c} r^*(y; P(z, p_0)) \\ -r_*(y; P(z, p_0)) \end{array} \right\} \geq c_1 \exp\{-(1 + \epsilon)u \log u\} \,,$$

where

$$u = \frac{\log y}{\log z} \quad \text{and} \quad c_1 = c_1(\epsilon)$$

is a positive constant depending at most on ϵ.

The proof is accomplished by studying the function

$$f(u) = \begin{cases} r(0, z^u; P(z)) & \text{if } u \geq 1 \\ -1 & \text{if } 0 \leq u < 1 \,. \end{cases}$$

Let

$$W(z) = \prod_{p \leq z} \left(1 - \frac{1}{p}\right) \,.$$

One considers

$$F(s) = \int_0^\infty f(u)e^{su}\,du \,,$$

the Laplace-transform of $f(u)$, where the integral converges absolutely in the half-plane $\text{Re } s < \frac{1}{2} \log z$, so that $F(s)$ is an analytic function in this half-plane.

By considering the identity for generating functions:

$$W(z)(\log z - s)F(s) = \prod_{p \leq z} \left(1 - \frac{1}{p^{1-s/\log z}}\right) \zeta \left(1 - \frac{s}{\log z}\right)$$

$$+ \frac{1}{s} W(z)(\log z - s) - \left(\frac{e^s}{z} - 1\right)$$

one may derive the following asymptotics:

For $s = \sigma + i\tau$, $1 \leq \sigma < \frac{1}{2} \log z$, $|\tau| \leq 10$, we have

$$F(s) = -\exp\left\{-\frac{e^s}{s} + O\left(\frac{e^\sigma}{\sigma}\right)\right\} + O(1). \tag{3.6}$$

By Plancherel's identity one may deduce

$$\int_0^\infty |f(u)e^{\sigma u}|^2 du = \frac{1}{2\pi} \int_{-\infty}^\infty |F(\sigma + i\tau)|^2 d\tau \geq \frac{1}{2\pi} \int_{\pi-1}^{\pi+1} |F(\sigma + i\tau)|^2 d\tau.$$

This together with the asymptotics (3.6) yields a lower bound for weighted averages of $|f(u)|^2$, from which the required oscillation result follows.

4   Irregularity Results for Primes in Arithmetic Progressions

Definition 4.1 Let $q \in \mathbb{N}$, $1 \leq a \leq q - 1$, $(a, q) = 1$. Then

$$\pi(x; q, a) = \sum_{\substack{p \leq x \\ p \equiv a \bmod q}} 1, \quad \psi(x; q, a) = \sum_{n \equiv a \bmod q} \Lambda(n).$$

As in the previous section we have to discuss the question: What is expected behavior?

For the discussion of this question see also the survey paper [9]. Much of the following will be an extension of the description given in [9].

Expectations for the behavior of $\pi(x; q, a)$ may be derived from Cramér's model. From Cramér's model one may derive the following:

$$\pi(x; q, a) = \frac{\pi(x)}{\phi(q)} + O\left(\left(\frac{x}{q}\right)^{1/2} \log(qx)\right),$$

uniformly in the range

$$q \le Q = \frac{x}{(\log x)^B}, \ (a, q) = 1 \text{ for any } B > 2 . \tag{4.1}$$

Another source for expectations on $\pi(x; q, a)$ may be the proof of regularity results. These results contain restrictions of the form $q \le Q = F(x)$ for the modulus q, similar to (4.1). In the sequel we list some results and conjectures about $\pi(x; q, a)$.

One generally expects the validity of the equidistribution result

$$\pi(x; q, a) \sim \frac{\pi(x)}{\phi(q)} . \tag{4.2}$$

It has been proven that (4.2) holds uniformly for:

(i) All $q \le \log^B x$ and all $(a, q) = 1$ for any fixed $B > 0$ (Siegel-Walfisz).
(ii) All $q \le \sqrt{x}/\log^{2+\epsilon} x$ and all $(a, q) = 1$, assuming the Generalized Riemann Hypothesis.
(iii) Almost all $q \le \sqrt{x}/\log^{2+\epsilon} x$ and all $(a, q) = 1$ (Bombieri-Vinogradov).
(iv) Almost all $q \le x^{1/2+o(1)}$ with $(a, q) = 1$ for fixed $a \ne 0$ (Bombieri-Friedlander-Iwaniec, Fouvry).
(v)

$$\text{Almost all } q \le x/\log^{2+\epsilon} x \text{ and almost all } (a, q) = 1 \tag{4.3}$$

(Barban-Davenport-Halberstam, Montgomery, Hooley).

Montgomery [16, 17] has conjectured that for any $\epsilon > 0$,

$$\left| \psi(x; q, a) - \frac{x}{\phi(q)} \right| \ll_\epsilon \left(\frac{x}{q} \right)^{\frac{1}{2}+\epsilon} \log x \tag{4.4}$$

holds uniformly in the range $q < x$ and $(a, q) = 1$, which would imply that (4.2) holds uniformly in the range $q < x/\log^{2+\epsilon} x$ and $(a, q) = 1$.

Elliott and Halberstam [3] have considered the estimate

$$\sum_{q \le Q} \phi(q) \max_{(a,q)=1} \left\{ \pi(x; q, a) - \frac{li(x)}{\phi(x)} \right\}^2 \ll_A \frac{x^2}{\log^A x} \tag{4.5}$$

and conjectured that for any $A > 1$, (4.6) holds if $Q < x/\log^{A+1} x$.

Friedlander and Granville [5] have established results, which refute the conjectures of Montgomery and of Elliott-Halberstam.

The central result of the paper [5] is Proposition 1: Let R be the set of integers q, that are free of prime factors $< \log q$.

Fix $B > 1$. There exist arbitrarily large values of Q, such that for any subset S of R, where each $q \in S$ is in the range

$$\frac{Q}{\log^{1/8} Q} < q < Q,$$

there exist values of a and x with

$$Q \log^B Q < x < 3Q \log^B Q,$$

for which

$$\sum_{\substack{q \in S \\ (a,q)=1}} \left| \psi(x; q, a) - \frac{x}{\phi(q)} \right| \gg_B \sum_{q \in S} \frac{x}{\phi(q)}. \qquad (4.6)$$

The proof of Proposition 1 again is obtained by combining oscillation results on sifting functions with the matrix method. The matrix this time is

$$\mathcal{M} = (rP + qs)_{\substack{P^{D-1} < r \leq 2P^{D-1} \\ 1 \leq s \leq P^D/q}}$$

with $P = P(z)$.

An important special case of Proposition 1 is obtained by the choice $S = \{q\}$, where q is a prime. It shows that for any fixed $B > 0$ the estimate (4.2) cannot hold uniformly for the range

$$q \leq Q = \frac{x}{\log^B x} \quad \text{and} \quad (a, q) = 1.$$

More specifically, there exist arithmetic progressions $a_\pm \pmod q$ and values

$$x_\pm \in [\phi(q) \log^B q, \ 2\phi(q) \log^B q],$$

such that

$$\pi(x_+; q, a_+) > (1 + \delta_B) \frac{\pi(x_+)}{\phi(q)}$$

and

$$\pi(x_-; q, a_-) < (1 - \delta_B) \frac{\pi(x_-)}{\phi(q)},$$

where $\delta_B > 0$ is a constant.

From Proposition 1 one can easily deduce Theorem 1 of [5]:

Fix $B > 1$. There exist arbitrarily large values of a and x for which

(i)

$$\sum_{\substack{T<q<4T \\ (q,a)=1}} \left| \psi(x; q, a) - \frac{x}{\phi(q)} \right| \gg_B \frac{x}{\log\log x}, \qquad (4.7)$$

where $T = x/4 \log^B x$.

(ii)

$$\sum_{\substack{q<x/\log^B x \\ (q,a)=1}} \left| \psi(x; q, a) - \frac{x}{\phi(q)} \right| \gg_B x . \qquad (4.8)$$

From these results one easily deduces that conjecture (4.3) is incorrect. The conjecture (4.5) fails, if $A > 2$. The theorem of Bombieri-Vinogradov (see (4.3), (iii)) is usually stated in the form:

$$\sum_{q<Q} \max_{(a,q)=1} \max_{y \le x} \left| \pi(y; q, a) - \frac{\pi(y)}{\phi(q)} \right| \ll \frac{x}{\log^A x}, \qquad (4.9)$$

where $Q = x^{1/2} (\log x)^{-B}$.

From (4.7), (4.8) it follows that (4.9) fails whenever $x(\log x)^{-B} \le Q \le x$.

The lower bound for the occurrence of irregularities in the estimates above may be improved, if one also allows variations in the residue class a.

In [5] the following results are proven by use of the matrix method:

Theorem 2 (of [5]) *Fix $B > 1$. There exist arbitrarily large values of y, such that for any value of Q in the range*

$$\frac{y}{\log^B y} < Q < \frac{y}{\log y \log\log y}$$

there exists $x (= x(Q))$ in the range

$$y < x < 3y$$

such that

$$\sum_{Q<q<2Q} \max_{(a,q)=1} \left| \psi(x; q, a) - \frac{x}{\phi(q)} \right| \gg_B \sum_{Q<q<2Q} \frac{x}{\phi(q)} .$$

Theorem 3 (of [5]) *Assume that the Generalized Riemann Hypothesis holds. For any positive-valued $g(x)$ tending to 0 as $x \to \infty$ let Q_g be the set of integers q with more than $\exp(g(p) \log_2 q)$ distinct prime factors. Fix $\epsilon > 0$ and $N > 1$. Then, for all sufficiently large $q \notin Q_g$ and for any value of y in the range*

$$q \log q < y < q \log^N q ,$$

we have

$$\left| \psi(x; q, a) - \frac{x}{\phi(q)} \right| \gg_{N,\epsilon,g} \frac{x}{\phi(q)}$$

for at least $q / \exp((\log q)^\epsilon)$ distinct values of $a \pmod{q}$, with $(a, q) = 1$, for some $x = x(a)$ in the range

$$y \leq x \leq 2y .$$

The range for the irregularities in the behavior of $\pi(x; q, a)$ has been considerably extended in the paper [7]. To establish the oscillation results for sifting functions one has to invest more effort than in the paper [11] described in the previous section. To measure the oscillations of the sifting function we give the following

Definition 4.2

$$\Phi_k(y, z) = \#\{n \leq y : p \mid n, \ p \nmid k \Rightarrow p \geq z\} = y \prod_{\substack{p<z \\ p \nmid k}} \left(1 - \frac{1}{p}\right) (1 + r_k(y, z)) ,$$

$$r(y, z) = r_1(y, z) .$$

A smoothed version of $r(y, z)$ is obtained from the generating Dirichlet series:

$$F(s) = F(z; s) = \prod_{p<z} \left(1 - \frac{1}{p^s}\right) \zeta(s) .$$

Let

$$\phi(s) = \phi(z; s) = \log F(z; s)$$

be the branch of the logarithm, such that $\phi(\sigma)$ is real for $0 < \sigma < 1$.
    The point $s_0 = s_0(y, z)$ is defined as the unique solution to the equation

$$\phi'(s_0) = -\log y = -u\mathcal{L}$$

satisfying

$$s_0 = \sigma_0 + it_0, \quad \frac{\epsilon}{2} < \sigma_0 < 1 - \frac{3}{\mathcal{L}}, \quad 0 < t_0 < \frac{2\pi}{\mathcal{L}}.$$

The function $\rho(y, z)$ is defined by

$$\rho(y, z) = e^{\gamma} F(z; s_0) \frac{y^{s_0-1}}{s_0\sqrt{\pi u/2}},$$

where u and $\mathcal{L}$ are defined by

$$u = \frac{\log y}{\log z}, \quad \mathcal{L} = \log z$$

and γ is the Euler-Masceroni constant.

Finally, the smoothed version of $r(y, z)$ is defined by

$$r(y, z; \lambda) = \frac{\sqrt{\lambda}}{\sqrt{2\pi}} \int_{-\infty}^{\infty} r(ye^v, z)e^{-\lambda v^2/2} dv,$$

where λ is a positive real parameter.

Let the set (R_α) be defined by

$$(R_\alpha) = \{z \geq z_0, \ y = z^u, \ u_0 \leq u \leq z^{1-\alpha}\},$$

where $\alpha > 0$ is fixed, z_0, u_0 sufficiently large, depending only on α.

The basic result on oscillation of sifting functions in [7] is:

Theorem C *Let $\epsilon > 0$ be fixed. For (y, z) in the range (R_ϵ) and $u \leq \mathcal{L}^{3/2-\epsilon}$, we have:*

$$r(y, z; \lambda) = \operatorname{Re} \rho(y, z) + O\left(\frac{|\rho(y, z)|}{\log u}\right).$$

Theorem C is proven by the saddle point method. The basic idea is to write $r(y, z)$ as a Perron integral over $F(s)y^{s-1}/s$ along a vertical line segment whose abscissa is Re s_0. The main contribution to the integral then comes from small neighborhoods of the points s_0 and $\bar{s}_0$ that are saddle points of the function $F(y)y^{s-1}$. The ranges from [5] for the irregularity result can be extended as follows: Eq. (4.9) fails with

$$Q = \frac{x}{\exp((A - \epsilon)(\log\log x)^2 / \log\log\log x)}.$$

It is also shown that (4.4) cannot hold for every integer a, prime to q, for

(i) Any $q \geq x \exp((\log x)^{1/5-\epsilon})$
(ii) Any $q \geq x \exp((\log x)^{1/3-\epsilon})$ that has less than $1.5 \log\log\log q$ distinct prime factors $< \log q$.
(iii) Almost any $q \in (y, 2y]$ for any $y \geq x / \exp\left((\log x)^{1/2-\epsilon}\right)$.

Under assumption of the Generalized Riemann Hypothesis, the values $1/5$ and $1/3$ in (i) and (ii) can be improved to $1/3$ and $1/2$, respectively.

One may also obtain results for fixed values of a if one considers the matrix

$$\mathcal{M} = (1 + j((R+i)P - 1) \underset{0 \leq j \leq R-1}{\scriptstyle 0 \leq i \leq y}$$

In [6], Corollary to the Theorem, the following result is obtained:
For any fixed integer $a \neq 0$ and real $N > 0$ the asymptotic formula

$$\pi(x; q, a) \sim \frac{\pi(x)}{\phi(q)}$$

cannot hold uniformly in the range

$$q < \frac{x}{(\log x)^N} \; .$$

It follows [7] that for almost all a with $0 < |a| < x/\log^B x$, there exists

$$q \in \left(\frac{x}{\log^B x}, \; \frac{2x}{\log^B x} \right)$$

coprime to a for which (4.2) does not hold.

We conclude this section with the detailed statement of two irregularity results of [7].

Theorem A1 *Let $\epsilon > 0$. Then for all $q \geq q_0(\epsilon)$ and all x satisfying*

$$\sum_{\substack{p|q \\ p < \log q}} 1 < \frac{3}{2} \log_3 q \, ,$$

and

$$q(\log q)^{1+\epsilon} < x \leq q \exp\left((\log q)^{5/11-\epsilon} \right) ,$$

there exist numbers $x_\pm$ with $x/2 < x_\pm \leq 2x$ and integers $a_\pm$ coprime with q, such that

$$\Delta(x_+; q, a_+) \geq y^{-(1+\epsilon)\delta_1(x,y)}$$

$$\Delta(x_-; q, a_-) \leq -y^{-(1+\epsilon)\delta_1(x,y)} ,$$

where $y = x/q$ *and*

$$\delta_1(x, y) = \frac{\log(\log y/\log_2 x)}{\log(\log x/\log y)}$$

and $\Delta(x; q, a)$ *is defined by*

$$\theta(x; q, a) = \sum_{\substack{p \le x \\ p \equiv a \bmod q}} \log p = \frac{x}{\phi(q)}(1 + \Delta(x; q, a))$$

Theorem A2 *Let* $\epsilon > 0$. *There exist* $N(\epsilon) > 0$ *and* $q_0 = q_0(\epsilon)$, *such that for any* $q > q_0$ *and any* x *with*

$$q(\log q)^{N(\epsilon)} < x \le q \exp\left((\log q)^{1/3}\right) ,$$

there exist numbers $x_\pm$ *with* $x/2 < x_\pm \le 2x$ *and integers* $a_\pm$ *coprime with* q, *such that*

$$\Delta(x_+; q, a_+) \ge \frac{1}{\log^5 x} y^{-(1+\epsilon)\delta_2(x,y)}$$

$$\Delta(x_-; q, a_-) \le -\frac{1}{\log^5 x} y^{-(1+\epsilon)\delta_2(x,y)} ,$$

where $y = x/q$ *and*

$$\delta_2(x, y) = 3 \frac{\log(\log y/\log_2 x)}{\log(\log x \ \log y)} .$$

5 Other Irregularity Results

Granville and Soundararajan [10] have developed a much more general setting for the results described in the previous sections. Some corollaries of their results are versions of these previous results, which in some respects are stronger and in other respects are weaker.

The basic idea of the paper [10] is an uncertainty principle on arithmetic sequences: arithmetic sequences satisfying certain conditions cannot be well-distributed simultaneously in short intervals and in arithmetic progressions. To describe their results we start with the following

Definition 5.1 Let $a(n)$ be an arithmetic function. For a given set of integers $\mathcal{A}$ let

$$\mathcal{A}(x) := \sum_{n \leq x} a(n) \,.$$

We will suppose that

$$\mathcal{A}_d(x) := \sum_{\substack{n \leq x \\ d \mid n}} a(n) \approx \frac{h(d)}{d} \, \mathcal{A}(x) \,,$$

where h is a multiplicative function, $(d, S) = 1$, where S is a finite set of "bad" primes,

$$\mathcal{A}(x; q, a) := \sum_{\substack{n \leq x \\ n \equiv a \bmod q}} a(n) \,.$$

For $(q, S) = 1$, let

$$\gamma_q := \prod_{p \mid q} \frac{p - 1}{p - h(n)}$$

and $f_q(a)$ a nonnegative multiplicative function of a for which $f_q(a) = f_q((a, q))$.

The authors now study the approximations

$$\mathcal{A}(x + y) - \mathcal{A}(x) \approx y \, \frac{\mathcal{A}(x)}{x}$$

$$\mathcal{A}(x; q, a) \approx \frac{f_q(a)}{q \gamma_q} \, \mathcal{A}(x) \,.$$

In their Proposition 2.2 they have reduced the problem to results on the oscillations of mean-values of arithmetic functions.

Proposition 5.2 *Suppose that* $(q, S) = 1$ *and define* $\Delta_q = \Delta_q(x)$ *by*

$$\Delta_q(x) := \max_{x/4 \leq y \leq x} \max_{a \bmod q} \frac{\left| \mathcal{A}(y; q, a) - \frac{f_q(a)}{q \gamma_q} \mathcal{A}(x) \right|}{\mathcal{A}(x)/\phi(q)} \,.$$

Let $q \leq \sqrt{x} \leq l \leq x/4$ *be positive coprime integers with* $(q, S) = (l, S) = 1$.

Then

$$\frac{q}{\phi(q)}\Delta_q(x) + \frac{l}{\phi(l)}\Delta_l(x) + x^{-1/2} \gg \left| \frac{1}{[x/2l]} \sum_{s \leq x/(2l)} \frac{f_q(s)}{\gamma_q} - 1 \right|.$$

The proof of Proposition 5.2 is accomplished by the matrix method, where the matrix $\mathcal{M}$ considered is defined as

$$\mathcal{M} = ((R+r)q + sl)_{\substack{1 \leq r \leq R \\ 1 \leq s \leq S}},$$

where

$$R := \left[\frac{x}{4q}\right] \geq \frac{\sqrt{x}}{5}$$

and

$$S := \left[\frac{x}{2l}\right] \leq \frac{\sqrt{x}}{2}.$$

This time—instead of simply counting entries of rows and columns—one is summing the values $a(n)$, n ranging over rows and columns of $\mathcal{M}$.

The oscillation results needed—again as described in previous sections—are derived by the study of generating functions:

$$F_q(s) := \sum_{n=1}^{+\infty} \frac{f_q(n)}{n^s} = \zeta(s)G_q(s),$$

where

$$G_q(s) := \prod_{p|q} \left(1 - \frac{1}{p^s}\right)\left(1 + \frac{f_q(p)}{p^s} + \frac{f_q(p^2)}{p^{2s}} + \cdots\right).$$

Setting

$$E(u) := \frac{1}{z^u} \sum_{n \leq z^u} (f_q(n) - G_q(1))$$

one may express the Laplace transform

$$I(s) = \int_0^{+\infty} e^{-su} E(u)\, du$$

by the generating function as follows:

$$I(s) = \frac{\zeta\left(1 + \frac{s}{\log z}\right)}{\log z + s} \left(G_q\left(1 + \frac{s}{\log z}\right) - G_q(1)\right).$$

We now discuss the corollaries about limitations to the equidistribution of primes.
We have

Theorem 5.3 (See [10]) *Let l be large and suppose that l has fewer than $(\log l)^{1-\epsilon}$ prime divisors below $\log l$. Suppose that*

$$(\log l)^{1+\epsilon} \le y \le \exp\left(\frac{\beta\sqrt{\log l}}{\sqrt{2\log\log l}}\right)$$

for a certain absolute constant $\beta > 0$, and set $x := ly$. Define for integers a coprime to l

$$\Delta(x; l, a) = \frac{\theta(x; l, a) - \frac{x}{\phi(l)}}{\frac{x}{\phi(l)}}.$$

There exist numbers $x_\pm$ in the interval

$$\left(x, xy^{D/\log(\log y/\log\log l)}\right)$$

and integers $a_\pm$ coprime to l, such that

$$\Delta(x_+; l, a_+) \ge y^{-\delta(l,y)} \quad \text{and} \quad \Delta(x_-; l, a_-) \le -y^{-\delta(l,y)}.$$

Here D is an absolute positive constant which depends only on ϵ, and

$$\delta(x, y) = \frac{1}{\log\log x}\left(\log\left(\frac{\log y}{\log\log x}\right) + \log\log\left(\frac{\log y}{\log\log x}\right) + O(1)\right).$$

One should compare this with [7], Theorem A1. The bound $y^{-(1+o(1))\tau/(1+\tau)}$ is weaker than the bound $y^{-\tau(1+o(1))}$, which is obtained in Theorem 5.3 for $y = \exp((\log l)^\tau)$. The constraint on the small primes dividing l is less restrictive in Theorem 5.3. However, the localization

$$x_\pm \in \left(x, xy^{D/\log(\log y/\log\log N)}\right)$$

in Theorem 5.3 is worse than the localization $x_\pm \in (x/2, 2x)$ in Theorem A1.
    The following Theorem contains no restriction on the small primes dividing l.
We have

Theorem 5.4 (See [10]) *Let l be large and suppose that*

$$(\log l)^{1+\epsilon} \leq y \leq \exp\left(\beta\sqrt{\log l}/\sqrt{\log\log l}\right)$$

for a certain absolute constant $\beta > 0$, *and set* $x := yl$.
   There exist numbers $x_\pm$ *in the interval*

$$\left(x, xy^{D/\log(\log y/\log\log x)}\right)$$

and integers $a_\pm$ *coprime to l, such that*

$$\Delta(x_+; l, a_+) \geq \frac{y^{-\delta_1(l,y)}}{\log\log\log l}$$

and

$$\Delta(x_-; l, a_-) \leq -\frac{y^{-\delta_1(l,y)}}{\log\log\log l},$$

where

$$\delta_1(x, y) = \frac{\log\log y + O(1)}{\log\log x}.$$

Here D is an absolute positive constant which depends only on ϵ.

   Theorem 5.4 should be compared with Theorem A2 of [7]. Our bound is

$$\gg y^{-\tau(1+o(1))}, \text{ if } y = \exp((\log l)^\tau) \text{ with } 0 < \tau < 1/2.$$

The corresponding result in Theorem A2 of [7] gives a weaker bound

$$y^{-(3+o(1))\tau/(1+\tau)}.$$

Again, the localization of the $x_\pm$ values is better in Theorem A2 of [7].
   The results of Granville and Soundararajan also include as a special case a previous result of Balog and Wooley [1] on sums of two squares in short intervals.
   Let

$$B := \frac{1}{\sqrt{2}} \prod_{p \equiv -1 \bmod 4} \left(1 - \frac{1}{p^2}\right)^{-1/2}.$$

Let the functions F, f be defined by

$$\begin{cases} F(s) = 2\sqrt{e^{\gamma}/\pi}\, s^{-1/2}, & \text{when } 0 < s \leq 1/2 \\ f(s) = 0, & \text{when } 0 < s \leq 1, \end{cases}$$

and

$$\begin{cases} (s^{1/2}F(s))' = \frac{1}{2}s^{-1/2}f(s-1), & \text{where } s > 2 \\ (s^{1/2}f(s))' = \frac{1}{2}s^{-1/2}F(s-1), & \text{where } s > 1. \end{cases}$$

Then there is the following result (Theorem 1 of [1]):

Let $N > 0$ be fixed. There are sequences of real numbers x^{+}, x^{-} tending to infinity, such that with $y := (\log x^{+})^{N}$, respectively $y = (\log x^{-})^{N}$

$$\text{Card}\{n : x^{+} < n \leq x^{+} + y, \, n = u^2 + v^2\} > \frac{By}{\sqrt{\log x^{+}}}(F(N) + o(1)),$$

respectively

$$\text{Card}\{n : x^{-} < n \leq x^{-} + y, \, n = u^2 + v^2\} < \frac{By}{\sqrt{\log x^{-}}}(f(N) + o(1)).$$

All the irregularity results discussed in the previous sections may be seen as statements on prime values of polynomials $f(t) = qt + a$, with $q = 1$ in the case of primes in short intervals. Polynomials of higher degree have been considered by Friedlander and Granville [6] as well as Nair and Perelli [18].

Nair and Perelli consider the polynomials $F_R(n) = n^d + RP$ and obtain irregularity results by considering the matrix

$$\mathcal{M} = (F_{R+j}(i))_{\substack{1 \leq i \leq y \\ 1 \leq j \leq R-1}}.$$

The matrix method has been applied to function fields by Thorne [22].

Let $\mathbb{F}_q$ be the finite field with q elements, and let $\mathbb{F}_q[t]$ denote the polynomial ring in one variable. The Riemann Hypothesis is known and results like the Prime Number Theorem may be proved with the strongest form

$$\pi(n) = \frac{q^n}{n} + O\left(\frac{q^{n/2}}{n}\right), \tag{5.1}$$

where $\pi(n)$ denotes the number of primes of degree n.

We come to the description of the results and the methods of proof.

Let $f \in \mathbb{F}_q[t]$ be a fixed polynomial and $n < \deg f$. The interval (f, n) is the set of polynomials $f + g$, where g ranges over all polynomials with $\deg g \leq n$.

The number of primes in this interval is denoted by $\pi(f, n)$. By (5.1) one expects that

$$\pi(f, n) \sim \frac{q^{n+1}}{\deg f}$$

for reasonably large n. This does not necessarily hold if n is sufficiently small in relation to $\deg f$.

One has Theorem 1.1. of [22]. For any fixed $\lambda_0 > 0$, we have:

$$\limsup_{k \to \infty} \; \sup_{\deg f = k} \; \frac{\pi(f, s(k))}{q^{s(k)+1}/k} > 1$$

and

$$\liminf_{k \to \infty} \; \inf_{\deg f = k} \; \frac{\pi(f, s(k))}{q^{s(k)+1}/k} < 1 \,,$$

where

$$s(k) := [\lambda_0 \log k] \,. \tag{5.2}$$

For the proof of (5.2) let

$$Q := Q(n) = \prod_{\deg p \leq n} p \,.$$

The author then considers the matrix $\mathcal{M} = (a_{ij})$ with entries

$$a_{ij} := g_i Q + h_j \,,$$

where g_i ranges over all monic polynomials of degree $2 \deg Q$ and h_j ranges over all polynomials of degree $\leq s$.

To estimate the primes in admissible columns one may apply the Prime Number Theorem in arithmetic progressions [20]

$$\pi(n; m, a) = \frac{1}{\phi(m)} \frac{q^n}{n} + O\left(\frac{q^{n/2}}{n}\right) \,,$$

where $(a, m) = 1$.

Here $\pi(n; m, a)$ denotes the number of primes of $\mathbb{F}_q[t]$ of degree n congruent to a modulo m and the Euler ϕ-function is defined by

$$\phi(m) = |(\mathbb{F}_q[t]/m\mathbb{F}_q[t])^*| \,.$$

Acknowledgements We would like to express our thanks to J. Friedlander and A. Granville for their useful comments.

A. Raigorodskii: I would like to acknowledge financial support from the grant NSh-6760.2018.1.

M. Th. Rassias: I would like to express my gratitude to the J. S. Latsis Foundation for their financial support provided under the auspices of my current "Latsis Foundation Senior Fellowship" position.

References

1. A. Balog, T.D. Wooley, Sums of two squares in short intervals. Can. J. Math. **52**(4), 673–694 (2000)
2. H. Cramér, On the order of magnitude of the difference between consecutive prime numbers. Acta Arith. **2**, 396–403 (1936)
3. P.D.T.A. Elliott, H. Halberstam, A conjecture in prime number theory. Symp. Math. **4**, 59–72 (1968)
4. P. Erdős, On the difference of consecutive primes. Q. J. Oxf. **6**, 124–128 (1935)
5. J. Friedlander, A. Granville, Limitations to the equidistribution of primes I. Ann. Math. **129**, 363–382 (1989)
6. J. Friedlander, A. Granville, Limitations to the equidistribution of primes IV. Proc. R. Soc. Lond. Ser. A **435**, 197–204 (1991)
7. J. Friedlander, A. Granville, A. Hildebrand, H. Maier, Oscillation theorems for primes in arithmetic progressions and for sifting functions. J. AMS **4**(1), 25–86 (1991)
8. P.X. Gallagher, A large sieve density estimate near $\sigma = 1$. Invent. Math. **11**, 329–339 (1970)
9. A. Granville, Unexpected irregularities in the distribution of prime numbers, in *Proceedings of the International Congress of Mathematicians*, Zürich (1994), pp. 388–399
10. A. Granville, K. Soundararajan, An uncertainty principle for arithmetic sequences. Ann. Math. **165**, 593–635 (2007)
11. A. Hildebrand, H. Maier, Irregularities in the distribution of primes in short intervals. J. Reine Angew. Math. **397**, 162–193 (1989)
12. H. Iwaniec, The sieve of Eratosthenes-Legendre. Ann. Sc. Norm. Super. Pisa Cl. Sci. **4**(4), 257–268 (1977)
13. H. Maier, Chains of large gaps between consecutive primes. Adv. Math. **39**, 257–269 (1981)
14. H. Maier, Primes in short intervals. Mich. Math. J. **32**(2), 221–225 (1985)
15. K. Monks, S. Peluse, L. Ye, Strings of special primes in arithmetic progressions (English summary). Arch. Math. (Basel) **101**(3), 219–234 (2013)
16. H.L. Montgomery, *Topics in Multiplicative Number Theory*. Lecture Notes in Mathematics, vol. 227 (Springer, New York, 1971)
17. H.L. Montgomery, Problems concerning prime numbers. Proc. Symp. Pure Math. **28**, 307–310 (1976)
18. A. Nair, A. Perelli, On the prime ideal theorem and irregularities in the distribution of primes. Duke Math. J. **77**, 1–20 (1995)
19. R.A. Rankin, The difference between consecutive prime numbers. J. Lond. Math. Soc. **13**, 242–247 (1938)
20. M. Rosen, *Number Theory in Function Fields*. Graduate Texts in Mathematics, vol. 210 (Springer, New York, 2002)
21. D.K.L. Shiu, Strings of congruent primes. J. Lond. Math. Soc. **61**(2), 359–373 (2000)
22. F. Thorne, Irregularities in the distribution of primes in function fields. J. Number Theory **128**(6), 1784–1794 (2008)

Sums of Values of Nonprincipal Characters over Shifted Primes

Zarullo Rakhmonov

Abstract For a nonprincipal character χ modulo D, when $x \geq D^{\frac{5}{6}+\varepsilon}$, $(l, D) = 1$, we prove a nontrivial estimate of the form $\sum_{n \leq x} \Lambda(n)\chi(n - l) \ll x \exp\left(-0.6\sqrt{\ln D}\right)$ for the sum of values of χ over a sequence of shifted primes.

1 Introduction

I.M. Vinogradov's method for estimating exponential sums over primes enabled him to solve a number of arithmetic problems with primes. One of these problems concerns the distribution of values of a nonprincipal character over sequences of shifted primes. In 1938, he proved the following: *If q is an odd prime, $(l, q) = 1$, and $\chi(a)$ is a nonprincipal character modulo q, then*

$$T_1(\chi) = \sum_{p \leq x} \chi(p - l) \ll x^{1+\varepsilon}\sqrt{\frac{1}{q} + \frac{q}{\sqrt[3]{x}}}.$$

(see [35]). In 1943, Vinogradov [36] improved this estimate by showing that

$$|T_1(\chi)| \ll x^{1+\varepsilon}\left(\sqrt{\frac{1}{q} + \frac{q}{x}} + x^{-\frac{1}{6}}\right). \tag{1}$$

For $x \gg q^{1+\varepsilon}$, this estimate is nontrivial and implies an *asymptotic formula for the number of quadratic residues (nonresidues) modulo q of the form $p - l$, $p \leq x$.*

A Goldbach number is a positive integer that can be represented as the sum of two odd primes. The problem of the distribution of such numbers in "short" arithmetic

Z. Rakhmonov (✉)
A. Juraev Institute of Mathematics, Academy of Sciences of the Republic of Tajikistan, Dushanbe, Tajikistan
e-mail: zarullo-r@rambler.ru

progressions has appeared in attempts to solve the binary Goldbach problem. The first result of a conditional nature belongs to Linnik [20]. Under the assumption of the Generalized Riemann Hypothesis, he obtained the upper bound

$$G(D, l) \leq D \ln^6 D,$$

where $G(D, l)$ is the smallest Goldbach number in the arithmetic progression

$$Dk + l, \qquad k = 0, 1, 2, \ldots.$$

This result was further refined by Prahar [23, 24] and Wang [41]. Under the same assumptions, they proved that

$$G(D, l) \leq D(\ln D)^{3+\varepsilon}.$$

In 1968 Jutila [6] proved an unconditional theorem. Using the estimate (1), he showed that if D is an odd prime number, then

$$G(D, l) \ll D^{\frac{11}{8}+\varepsilon}.$$

Then I. M. Vinogradov obtained a nontrivial estimate for $T_1(\chi)$ in the case when $x \geq q^{0.75+\varepsilon}$, where q is a prime (see [37, 38, 40]). This result was unexpected. The point is that $T_1(\chi)$ can be expressed as a sum over the zeros of the corresponding Dirichlet L-function. Then, assuming that the Generalized Riemann hypothesis is valid for $T_1(\chi)$, one obtains a nontrivial estimate but only for $x \geq q^{1+\varepsilon}$.

It seemed that one got what was impossible. In 1973, in this connection Linnik [21] wrote, *"Vinogradov's investigations into the asymptotic behavior of Dirichlet characters are of great importance. As early as 1952, he obtained an estimate for the sum of the Dirichlet characters of shifted primes $T_1(\chi)$, which gave a degree of decrease relative to x as soon as $x > q^{0.75+\varepsilon}$, where q is the modulus of the character. This has fundamental significance, since it is deeper than what can be obtained by direct application of the Generalized Riemann hypothesis, and in this direction seems to carry a deeper truth than that hypothesis (if the hypothesis is valid). Recently A. A. Karatsuba has been able to improve this estimate."*

In 1968, Karatsuba developed a method that allowed him to obtain a nontrivial estimate for short character sums in finite fields of fixed degree [7, 8, 18]. In 1970, he improved this method and, combining it with Vinogradov's method, established the following statement [9, 10, 18]: *If q is a prime, $\chi(a)$ is a nonprincipal character modulo q, and $x \geq q^{\frac{1}{2}+\varepsilon}$, then*

$$T_1(\chi) \ll x q^{-\frac{1}{1024}\varepsilon^2}.$$

Karatsuba used these estimates to find asymptotic formulas for the number of quadratic residues and nonresidues of the form $p+k$ and for the number of products of the form $p(p' + k)$ in an arithmetic progression with increasing difference [11] (see also [12–16, 18]).

The present author generalized estimate (1) to the case of a composite modulus and proved the following statement [25, 26, 28]: *Let D be a sufficiently large positive integer, χ be a nonprincipal character modulo D, χ_q be the primitive character generated by χ_q, and q_1 be the product of primes that divide D but do not divide q; then*

$$T_1(\chi) \le x \ln^5 x \left(\sqrt{\frac{1}{q} + \frac{q}{x}\tau^2(q_1)} + x^{-\frac{1}{6}}\tau(q_1) \right) \tau(q).$$

Applying this estimate, he proved that the estimate

$$G(D, l) \ll D^{c+\varepsilon},$$

holds for a sufficiently large odd integer D (see [25, 27]), where ε is arbitrarily small positive constant, c is the lower bound of such a's, for which the inequality

$$\sum_{\chi \bmod D} N(\alpha, T, \chi) \ll (DT)^{2a(1-\alpha)} (\ln DT)^A .$$

holds for some constant $A > 2$. In fact, Huxley's "density" theorem [5] allows one to obtain the estimate $c \le \frac{6}{5}$ by taking $A = 14$.

In 2010, Friedlander et al. [3] showed that for a composite q the sum $T(\chi_q)$ can be estimated nontrivially when the length x of the sum is of smaller order than q. They proved the following: *For a primitive character χ_q and any $\varepsilon > 0$, there exists a $\delta > 0$ such that the following estimate holds for all $x \ge q^{\frac{8}{9}+\varepsilon}$:*

$$T_1(\chi_q) \ll xq^{-\delta}. \qquad (FGSh)$$

In 2013, the present author proved the following theorem (see [29–31]): *If q is a sufficiently large positive integer, χ_q is a primitive character modulo q, $(l, q) = 1$, ε is an arbitrarily small fixed positive number, and $x \ge q^{\frac{5}{6}+\varepsilon}$, then*

$$T_1(\chi_q) \ll x \exp\left(-\sqrt{\ln q}\right).$$

In 2017 Bryce Kerr in [19] obtained an estimate (FGSh), i.e. his estimate for $T_1(\chi_q)$ is nontrivial and gives a power saving factor when $x \ge q^{\frac{5}{6}+o(1)}$

In 2017, the author (see [32, 33]) proved the following theorem: *Let D be a sufficiently large positive integer, χ be a nonprincipal character modulo D, χ_q be the primitive character modulo q generated by χ, q be a cube-free number, $(l, D) = 1$, and ε an arbitrarily small positive constant. Then, for $x \ge D^{\frac{1}{2}+\varepsilon}$, we have*

$$T_1(\chi) \ll x \exp\left(-0.6\sqrt{\ln D}\right).$$

As was noted above, nontrivial estimates of the sum $T_1(\chi)$, where χ is a nonprincipal character modulo D, where D is a prime number, have been applied in problems of finding the least Goldbach number and the distribution of products of shifted primes in short arithmetic progressions. To solve these problems for composite modulus D, along with nontrivial estimates of the sum $T_1(\chi)$ for primitive characters, similar estimates are required for imprimitive characters. Therefore, we obviously must consider a nontrivial estimation problem for the sum $T_1(\chi)$, where χ is a nonprincipal character modulo a composite number D.

In this paper we obtain a nontrivial estimate for the sum $T(\chi)$ for all nonprincipal characters modulo a composite number. Let us state the main result.

Theorem 1 *Suppose that D is a sufficiently large positive integer, χ is a nonprincipal character modulo D, $(l, D) = 1$, ε is an arbitrarily small positive constant. Then for $x \geq D^{\frac{5}{6}+\varepsilon}$, we have*

$$T(\chi) = \sum_{n \leq x} \Lambda(n)\chi(n - l) \ll x \exp\left(-0.6\sqrt{\ln D}\right),$$

where the constant under the sign $\ll$ depends only on ε.

Notations In what follows, we shall always assume that D is a sufficiently large natural number, x, l are natural numbers, $(l, D) = 1$, χ is a non-principal character modulo D, χ_q is a primitive character modulo q, generated by the character χ, q_1 is product of prime numbers dividing D but not dividing q, hence $(q, q_1) = 1$, ν is divisor of q_1, $\mathscr{L} = \ln D$, $\mathscr{L}_q = \ln q$, $\delta \leq 10^{-4}$ is a fixed positive number, c is a fixed positive number, possibly not the same at all times, $\omega(q)$ is the number of distinct prime divisors of q, which admits well-known estimate

$$\omega(q) \leq \frac{c_\omega \mathscr{L}_q}{\ln \mathscr{L}_q}, \tag{2}$$

.

In what follows, we shall frequently use the following well-known lemmas.

Lemma 1 *Let $f(n)$ be an arbitrary complex-valued function, $u_1 \leq x$, $r \geq 1$,*

$$C_r^k = \frac{r!}{k!(r - k)!}, \quad \lambda(n) = \sum_{d|n,\ d \leq u_1} \mu(d).$$

Then the following identity holds:

$$\sum_{n \leq x} \Lambda(n) f(n)$$

$$= \sum_{k=1}^{r} (-1)^{k-1} C_r^k \sum_{m_1 \leq u_1} \mu(m_1) \cdots \sum_{\substack{m_k \leq u_1 \\ m_1 \cdots m_k n_1 \cdots n_k \leq x}} \mu(m_k) \sum_{n_1} \cdots \sum_{n_k} \ln n_1 f(m_1 n_1 \cdots m_k m_k)$$

$$+ (-1)^r \sum_{n_1 > u_1} \lambda(n_1) \cdots \sum_{\substack{n_r > u_1 \\ n_1 \cdots n_r m \leq x}} \lambda(n_r) \sum_{m} \Lambda(m) f(n_1 \cdots n_r m).$$

Proof Proved in [34] using Heath-Brown identity (see [4]).

Lemma 2 *Let* $F(x, z, b)$ *be the number of positive integers less than* x *and relatively prime to* b, $b \le x$ *and having only prime divisors less than* z, $\ln x \le z \le x^{\frac{1}{e}}$, $\alpha = \ln z / \ln x$; *then for some* $|\theta| \le 1$, *we have*

$$F(x, z, b) \ll x \prod_{p|b}\left(1 - \frac{1}{p}\right) \exp\left(-\frac{1}{\alpha}\left(\ln\frac{1}{\alpha} + \ln\ln\frac{1}{\alpha}\right) + \frac{1}{\alpha} + \frac{20\theta}{\alpha \ln \frac{1}{\alpha}}\right).$$

Proof [39].

Lemma 3 *Let* r *be an arbitrary fixed natural number,* Z *is a natural number,* q *is a square-free number or* $r = 2$. *Then the following relations hold:*

$$\sum_{\lambda=0}^{q-1}\left|\sum_{z=1}^{Z}\chi_q(\lambda + z)\right|^{2r} \ll Z^r q + Z^{2r} q^{\frac{1}{2}+\delta},$$

where the constant under the sign $\ll$ *depends only on* r *and* δ.

Proof [1].

Lemma 4 *For an arbitrary natural number* $Z \le q^{\frac{1}{6}}$ *the following relations hold:*

$$\sum_{z_1,\dots,z_6=1}^{Z}\left|\sum_{\lambda=0}^{q-1}\chi_q\left(\frac{(\lambda + z_1)(\lambda + z_2)(\lambda + z_3)}{(\lambda + z_4)(\lambda + z_5)(\lambda + z_6)}\right)\right| \ll Z^3 q^{1+\delta}.$$

Proof [2].

Lemma 5 *The following asymptotic formula holds for all positive integers* q *and* U:

$$\left|\sum_{\substack{u=1 \\ (u,q)=1}}^{U} 1 - \frac{\varphi(q)}{q}U\right| \le 2^{\omega(q)}.$$

Proof We have

$$\sum_{\substack{u=1 \\ (u,q)=1}}^{U} 1 = \sum_{u=1}^{U}\sum_{d|(u,q)}\mu(d) = \sum_{d|q}\mu(d)\left[\frac{U}{d}\right] = U\sum_{d|q}\frac{\mu(d)}{d} - \sum_{d\backslash q}\mu(d)\left\{\frac{U}{d}\right\}.$$

Consequently

$$\left|\sum_{\substack{u=1 \\ (u,q)=1}}^{U} 1 - \frac{\varphi(q)}{q}U\right| = \left|\sum_{d\backslash q}\mu(d)\left\{\frac{U}{d}\right\}\right| \le \sum_{d|q}\mu^2(d) = \prod_{p|q}(1 + \mu^2(p)) = 2^{\omega(q)}.$$

Lemma 6 *For $x \geq 2$ we have*

$$\sum_{n \leq x} \tau_r^k(n) \ll x\,(\ln x)^{r^k-1}, \qquad k = 1, 2.$$

Proof [22].

2   Auxiliary Lemmas

Lemma 7 *Let $q|D$, then*

$$\sum_{\substack{d|q \\ d > \exp(\sqrt{2\mathscr{L}})}} \frac{\mu^2(d)}{d} \ll \exp\left(-0.7\sqrt{\mathscr{L}}\right).$$

Proof After dividing the interval of summation into intervals of the form $M < d \leq 2M$, we obtain less than $\mathscr{L}_q$ sums $S(M)$ of the form

$$S(M) = \sum_{\substack{d|q \\ M < d \leq 2M}} \frac{\mu^2(d)}{d} \ll M^{-1} \sum_{\substack{d|q \\ d \leq 2M}} 1.$$

Let $q = p_1^{\alpha_1} p_2^{\alpha_2} \ldots p_t^{\alpha_t}$ be the canonical decomposition of q into prime factors and let r_i denote ith prime number. Obviously, there exists such k that

$$q' = r_1 r_2 \ldots r_k \leq q < r_1 r_2 \ldots r_k r_{k+1}, \qquad k \geq t.$$

By the prime number theorem we have

$$\ln q' = \sum_{i \leq k} \ln r_i = \sum_{p \leq r_k} \ln p > \frac{r_k}{2},$$

thus

$$r_k < 2\ln q' \leq 2\mathscr{L}_q.$$

Let $q'' = r_1^{\alpha_1} r_2^{\alpha_2} \ldots r_t^{\alpha_t}$. Due to $q'' \leq q$ and $r_t \leq r_k$, we have

$$S(M) \ll M^{-1} \sum_{\substack{d|q \\ d \leq 2M}} 1 \leq M^{-1} \sum_{\substack{d|q'' \\ d \leq 2M}} 1.$$

Prime factors of numbers d, $d|q''$ satisfy the condition $r_j \leq r_t$. As $r_t < 2\mathcal{L}_q$, the latter sum does not exceed the number of such positive integers that are less than $2M$, and have only prime divisors less than $2\mathcal{L}_q$, i.e.

$$S(M) \ll M^{-1} \sum_{\substack{d|q'' \\ d \leq 2M}} 1 \leq M^{-1}F(2M, 2\mathcal{L}_q, 1) \leq M^{-1}F(2M, 2\mathcal{L}, 1).$$

Applying Lemma 2 with

$$x = 2M, \qquad b = 1, \qquad z = 2\mathcal{L}, \qquad \alpha = \frac{\ln z}{\ln x} = \frac{\ln 2\mathcal{L}}{\ln 2M},$$

we have

$$S(M) \ll \exp\left(-\frac{\ln 2M}{\ln 2\mathcal{L}}(\ln \ln 2M - \mathcal{B})\right),$$

$$\mathcal{B} = \ln \ln 2\mathcal{L} - \ln \ln \frac{\ln 2M}{\ln 2\mathcal{L}} + 1 + 2\theta \left(\ln \frac{\ln 2M}{\ln 2\mathcal{L}}\right)^{-1}.$$

The condition $2M > \exp(\sqrt{2\mathcal{L}})$ implies that $2\mathcal{L} < (\ln 2M)^2$. Therefore,

$$\mathcal{B} \leq \ln \ln (\ln 2M)^2 - \ln \ln \frac{\ln 2M}{\ln (\ln 2M)^2} + 1 + 2\left(\ln \frac{\ln 2M}{\ln (\ln 2M)^2}\right)^{-1}$$

$$= 1 + \ln 2 - \ln\left(1 - \frac{\ln 2 + \ln \ln \ln 2M}{\ln \ln 2M}\right) + \frac{2}{\ln \ln 2M}\left(1 - \frac{\ln 2 + \ln \ln \ln 2M}{\ln \ln 2M}\right)^{-1}$$

$$= 1 + \ln 2 + O\left(\frac{\ln 2 + \ln \ln \ln 2M}{\ln \ln 2M}\right) < 3 < 0.002 \ln \ln 2M.$$

Consequently,

$$S(M) \ll \exp\left(-\frac{\ln 2M}{\ln 2\mathcal{L}}(\ln \ln 2M - \mathcal{B})\right) \ll \exp\left(-0.998 \cdot \frac{\ln 2M \ln \ln 2M}{\ln 2\mathcal{L}}\right).$$

The condition $2M > \exp(\sqrt{2\mathcal{L}})$ implies that $\ln 2M > \sqrt{2\mathcal{L}}$, $\ln \ln 2M > 0.5 \ln 2\mathcal{L}$, therefore

$$S(M) \ll \exp\left(-0.499\sqrt{2}\sqrt{\mathcal{L}}\right) \ll \mathcal{L}^{-1} \cdot \exp\left(-0.7\sqrt{\mathcal{L}}\right).$$

This completes the proof of Lemma.

Lemma 8 *Let K denote the number of solutions of the congruence*

$$(nd + \eta k)y \equiv (n_1 d + \eta k)y_1 \pmod{q},$$

$$M < n, n_1 \le M + N, \quad 1 \le y, y_1 \le Y, \quad (y, q) = 1, \quad (y_1, q) = 1,$$

where $(\eta, q) = (k, d) = 1$, d divides q, $2NY < q$, $d < Y$ and $\rho(qd^{-1}, Y)$ denotes the number of divisors β of the integer qd^{-1}, satisfying the conditions $qY^{-1} \le \beta < qd^{-1}$ and $(\beta, d) = 1$. Then the following relation holds:

$$K \le NY + \frac{2Y^2}{d} + \frac{2Y^2}{d}\rho(qd^{-1}, Y) + \frac{2(NY)^{1+\delta}}{d},$$

where δ is an arbitrarily small positive constant.

Proof Dividing both sides of the congruence by y when $y = y_1$, we have

$$nd + \eta k \equiv n_1 d + \eta k \pmod{q}, \quad M < n, n_1 \le M + N, \ 1 \le y \le Y, \ (y, q) = 1,$$

or

$$nd \equiv n_1 d \pmod{q}, \quad M < n, n_1 \le M + N, \quad 1 \le y \le Y, \quad (y, q) = 1.$$

Let us divide both sides of the congruence and modulus by d, we have

$$n - n_1 \equiv 0 \pmod{qd^{-1}}, \quad M < n, n_1 \le M + N, \quad 1 \le y \le Y, \quad (y, q) = 1.$$

It follows from the conditions $|n - n_1| < N$ and $2N \le qY^{-1} < qd^{-1}$ that the latter congruence becomes an equality

$$n - n_1 = 0, \quad M < n, n_1 \le M + N, \quad 1 \le y \le Y, \quad (y, q) = 1,$$

i.e., $n = n_1$ if $y = y_1$. Thus, we obtain

$$K \le NY + 2\kappa, \tag{3}$$

where κ is the number of solutions of the congruence

$$(nd + \eta k)y \equiv (n_1 d + \eta k)y_1 \pmod{q},$$

$$M < n, n_1 \le M + N, \quad 1 \le y < y_1 \le Y, \quad (y, q) = 1, \quad (y_1, q) = 1,$$

or the congruence

$$(ny - n_1 y_1)d \equiv \eta k(y_1 - y) \pmod{q}, \tag{4}$$

$$M < n, n_1 \le M + N, \quad 1 \le y < y_1 \le Y, \quad (y, q) = 1, \quad (y_1, q) = 1.$$

The left-hand side of congruence (4) and its modulus is divisibly by d. Therefore, its right-hand side, that is, $\eta k(y_1 - y)$, is also divisible by d. As ηk is coprime with d, the number d must divide $y_1 - y$, i.e. $y_1 - y \equiv 0 \pmod d$, or equivalently, $y_1 = y + td$. Therefore, the congruence (4) can be expressed in the form

$$(ny - n_1(y + td))d \equiv \eta ktd \pmod q,$$

$$M < n, n_1 \leq M + N, \quad 1 \leq y < y + td \leq Y, \quad (y, q) = 1, \quad (y + td, q) = 1.$$

Dividing both sides of the congruence and its modulus by d, we obtain

$$(n - n_1)y \equiv (n_1 d + \eta k)t \pmod{qd^{-1}}, \tag{5}$$

$$M < n, n_1 \leq M + N, \quad 1 \leq y < y + td \leq Y, \quad (y, q) = 1, \quad (y + td, q) = 1.$$

Let us split the set of solutions of (5)

$$\kappa = \kappa_1 + \kappa_2 + \kappa_3, \tag{6}$$

into three subsets, where κ_1, κ_2, and κ_3 denote the number of solutions of the congruence (5) having, respectively, the following properties:

1. $n_1 d + \eta k \equiv 0 \pmod{qd^{-1}}$;
2. $(n_1 d + \eta k)t \equiv 0 \pmod{qd^{-1}}$ and $n_1 d + \eta k \not\equiv 0 \pmod{qd^{-1}}$;
3. $(n_1 d + \eta k)t \not\equiv 0 \pmod{qd^{-1}}$.

Estimate of κ_1. The congruence $n_1 d + \eta k \equiv 0 \pmod{qd^{-1}}$ has no solution for $(d, qd^{-1}) > 1$, and for $(d, qd^{-1}) = 1$ has no more than one solution $n_1 = n_1^*$, $(n_1^*, qd^{-1}) = 1$ as $2N < qd^{-1}$, i.e., N—the length of the interval of possible values for n_1—is less than the modulus of congruence. For $n_1 = n_1^*$ the congruence (5) becomes

$$(n - n_1^*)y \equiv 0 \pmod{qd^{-1}}, \quad M < n \leq M + N,$$

$$1 \leq y < y + td \leq Y, \quad (y, q) = 1, \quad (y + td, q) = 1,$$

and for fixed y and t has a single solution $n = n_1^*$, therefore

$$\kappa_1 \leq Y \left(\frac{Y}{d} + 1 \right) \leq \frac{2Y^2}{d}.$$

Estimate of κ_2. Using condition of the case 2, we represent the congruence (5) as a system of congruences

$$(n_1 - n)y \equiv 0 \pmod{qd^{-1}}, \tag{7}$$

$$(n_1 d + \eta k)t \equiv 0 \pmod{qd^{-1}},$$

satisfying conditions

$$n_1 d + \eta k \not\equiv 0 \,(\mathrm{mod}\ qd^{-1}), \quad M < n, n_1 \leq M + N,$$

$$1 \leq y < y + td \leq Y, \quad (y, q) = (y + td, q) = 1.$$

It follows from the conditions $(y, q) = 1$, $|n - n_1| < N$ and $2N \leq qd^{-1}$ that the first congruence in (7) is equivalent to the equality $n_1 = n$, therefore the number of solutions of the system (7) equals the number of solutions of the system

$$(nd + \eta k)t \equiv 0 \ (\mathrm{mod}\ qd^{-1}), \quad nd + \eta k \not\equiv 0 \,(\mathrm{mod}\ qd^{-1}), \tag{8}$$

$$M < n \leq M + N, \quad 1 \leq y < y + td \leq Y, \quad (y, q) = 1, \quad (y + td, q) = 1.$$

The product of numbers $nd + \eta k$ and t is divisible by qd^{-1}, but $nd + \eta k$ is not divisible by qd^{-1}, therefore for every solution of the congruence (8) there exists the divisor β of the number ad^{-1}, such that $\beta < qd^{-1}$ and

$$nd + \eta k \equiv 0 \ (\mathrm{mod}\ \beta), \qquad t \equiv 0 \,(\mathrm{mod}\ q(d\beta)^{-1}).$$

It follows from the condition $(d, \eta k) = 1$ that the congruence $nd + \eta k \equiv 0 \,(\mathrm{mod}\ \beta)$ has a solution only for $(\beta, d) = 1$. For $(\beta, d) = 1$ let us denote by $\kappa_2(\beta)$ the number of solutions of the system of congruences

$$n \equiv \eta d_\beta^{-1} \,(\mathrm{mod}\ \beta), \quad M < n \leq M + N, \quad dd_\beta^{-1} \equiv 1 \,(\mathrm{mod}\ \beta)$$

$$t \equiv 0 \,(\mathrm{mod}\ q(d\beta)^{-1}), \quad 1 \leq y < y + td \leq Y, \quad (y, q) = 1, \quad (y + td, q) = 1,$$

or the number of solutions of the congruence

$$n \equiv \eta d_\beta^{-1} \,(\mathrm{mod}\ \beta), \quad M < n \leq M + N, \quad dd_\beta^{-1} \equiv 1 \,(\mathrm{mod}\ \beta)$$

$$1 \leq y < y + tq\beta^{-1} \leq Y, \quad (y, q) = 1, \quad (y + tq\beta^{-1}, q) = 1.$$

The limits of the variables y and t in the last congruence can be expressed as

$$1 \leq y < Y, \quad 1 \leq t \leq \frac{Y - y}{q\beta^{-1}}, \quad (y, q) = 1, \quad (y + tq\beta^{-1}, q) = 1. \tag{9}$$

When $y > Y - q\beta^{-1}$, the upper bound of the variable t is smaller than the lower bound, therefore the region (9) can be represented as

$$1 \leq y \leq Y - q\beta^{-1}, \quad 1 \leq t \leq \frac{Y - y}{q\beta^{-1}}, \quad (y, q) = 1, \quad (y + tq\beta^{-1}, q) = 1.$$

In its turn, if $\beta \leq qY^{-1}$, then $Y - q\beta^{-1}$—the upper bound of y is smaller than the the lower bound. Therefore, $\kappa_2(\beta) = 0$ when $\beta \leq qY^{-1}$, while for $qY^{-1} < \beta < qd^{-1}$ we have the following estimate for $\kappa_2(\beta)$:

$$\kappa_2(\beta) \leq \left(\frac{N}{\beta} + 1\right) \sum_{\substack{1 \leq y < Y - q\beta^{-1} \\ (y,q)=1}} \left[\frac{Y - y}{q\beta^{-1}}\right] \leq \left(\frac{NY}{q} + \frac{Y\beta}{q}\right) \sum_{\substack{1 \leq y < Y - q\beta^{-1} \\ (y,q)=1}} 1.$$

Using the relations $2NY < q$, $\beta < qd^{-1}$ and $d < Y$, we find that

$$\kappa_2(\beta) \leq \left(1 + \frac{Y}{d}\right) \sum_{\substack{1 \leq y < Y - q\beta^{-1} \\ (y,q)=1}} 1 \leq \frac{2Y}{d} \sum_{\substack{1 \leq y < Y - q\beta^{-1} \\ (y,q)=1}} 1 < \frac{2Y^2}{d}.$$

Summing the last inequality over divisors β of the number qd^{-1}, satisfying the conditions $qY^{-1} \leq \beta < qd^{-1}$ and $(\beta, d) = 1$ and denoting the number of such divisors by $\rho(qd^{-1}, Y)$, we obtain

$$\kappa_2 \leq \sum_{\substack{(\beta,d)=1, \ \beta|qd^{-1} \\ qY^{-1} \leq \beta < qd^{-1}}} \kappa_2(\beta) \leq \frac{2Y^2}{d} \rho(qd^{-1}, Y).$$

Estimate of κ_3. Let us recall that κ_3 denotes the number of solutions of the congruence

$$(n_1 - n)y \equiv (n_1 d + \eta k)t \pmod{qd^{-1}},$$

satisfying the conditions

$$(n_1 d + \eta k)t \not\equiv 0 \pmod{qd^{-1}}, \quad M < n, n_1 \leq M + N,$$
$$1 \leq y < y + td \leq Y, \quad (y, q) = 1, \quad (y + td, q) = 1.$$

Fox a fixed pair (n_1^*, t^*) we denote by $\kappa_3(\lambda)$ the number of solutions of the congruence

$$(n - n_1^*)y \equiv \lambda \pmod{qd^{-1}}, \ M < n \leq M + N, \ 1 \leq y < y + t^*d \leq Y, \ (y, q) = 1, \tag{10}$$

where $0 < |\lambda| \leq q/2d$ is the absolute least residue of the number $(n_1^* d + \eta k)t^*$ modulo qd^{-1}. Using the bounds of variables n, n_1 and y and condition $2NY < q$, we find that

$$0 < |(n - n_1^*)y| < NY < \frac{q}{2}.$$

It follows that the inequality (10) becomes an equality

$$(n - n_1^*)y = \lambda, \quad M < n \leq M + N, \quad 1 \leq y < y + t^*d \leq Y, \quad (y, q) = 1, \tag{11}$$

where the parameter λ satisfies

$$1 \leq |\lambda| < NY.$$

Thus, for a fixed pair (n_1^*, t^*) the number of solutions $\kappa_3(\lambda)$ of the congruence (10) equals the number of solutions of the Eq. (11), and satisfies the following inequality

$$\kappa_3(\lambda) \leq \tau(|\lambda|) \leq 0.5 \, (NY)^\delta.$$

The number of possible pairs (n_1^*, t^*) does not exceed $N\left(Yd^{-1} + 1\right)$. Consequently,

$$\kappa_3 \leq N\left(\frac{Y}{d} + 1\right) \cdot 0.5(NY)^\delta \leq \frac{(NY)^{1+\delta}}{d}.$$

Inserting the estimates for κ_1, κ_2, and κ_3 into (6), and then into (3), we find that

$$K \leq NY + 2(\kappa_1 + \kappa_2 + \kappa_3) \leq NY + \frac{2Y^2}{d} + \frac{2Y^2}{d}\rho(qd^{-1}, Y) + \frac{2(NY)^{1+\delta}}{d}.$$

This completes the proof of Lemma.

3   Estimates of Short Character Sums

Lemma 9 *Let M, N, d, k, and η be natural numbers, $(\eta, q) = (d, k) = 1$, and*

$$S = \sum_{M-N < n \leq M} \chi_q(nd + \eta k).$$

Then for $N < q^{\frac{7}{12}}d^{-\frac{1}{2}}$, $D^{\frac{1}{2}} \leq q \leq D$ and $d \leq \exp(\sqrt{2\mathscr{L}})$, the following estimate holds:

$$|S| \leq N^{\frac{2}{3}}q^{\frac{1}{9} + \frac{\delta}{2}}d^{\frac{2}{3}}. \tag{12}$$

Proof The inequality (12) for the sum S will be proved by induction on N. For $N \leq q^{\frac{1}{3}}$ and $d \leq \exp(\sqrt{2\mathscr{L}})$ the right side of (12) admits an estimate

$$N^{\frac{2}{3}}q^{\frac{1}{9}+\frac{\delta}{2}}d^{\frac{2}{3}} \geq N^{\frac{2}{3}}q^{\frac{1}{9}+\frac{\delta}{2}} > N^{\frac{2}{3}}q^{\frac{1}{9}} \geq N^{\frac{2}{3}}\left(N^3\right)^{\frac{1}{9}} = N,$$

i.e., the inequality (12) is trivial and we take it as an induction basis.

Further, we assume that

$$N > q^{\frac{1}{3}}, \qquad d \leq \exp(\sqrt{2\mathscr{L}}).$$

Shifting the interval of summation by h, $1 \leq h \leq H < N$, in the sum S, we obtain

$$S = \sum_{M-N < n \leq M} \chi_q((n+h)d + \eta k) + \sum_{M-N < n \leq M-N+h} \chi_q(nd + \eta k)$$
$$- \sum_{M < n \leq M+h} \chi_q(nd + \eta k).$$

Estimating the last two sums using induction hypothesis, we have

$$|S| \leq \left| \sum_{M-N < n \leq M} \chi_q((n+h)d + \eta k) \right| + 2H^{\frac{2}{3}}q^{\frac{1}{9}+\frac{\delta}{2}}d^{\frac{2}{3}},$$

Setting $h = yz$ in this inequality and summing it over y and z satisfying

$$1 \leq y \leq Y, \quad (y,q) = 1, \quad 1 \leq z \leq Z, \quad Y = \left[0.5Nq^{-\frac{1}{6}}d\right], \quad Z = \left[0.5q^{\frac{1}{6}}d^{-1}\right],$$

we obtain

$$|S| \leq (Y_q Z)^{-1} \left| \sum_{\substack{1 \leq y \leq Y \\ (y,q)=1}} \sum_{1 \leq z \leq Z} \sum_{M-N < n \leq M} \chi_q((n+yz)d + \eta k) \right| + 2(YZ)^{\frac{2}{3}}q^{\frac{1}{9}+\frac{\delta}{2}}d^{\frac{2}{3}},$$

where Y_q denotes the number of integers $y \in [1, Y]$ coprime to q. Determining the number y^{-1} from the congruence $yy^{-1} \equiv 1 \pmod{q}$, we find that

$$|S| \leq (Y_q Z)^{-1} \sum_{M-N < n \leq M} \sum_{\substack{1 \leq y \leq Y \\ (y,q)=1}} \left| \sum_{1 \leq z \leq Z} \chi_q((nd + \eta k)y^{-1} + zd) \right|$$
$$+ 2^{-\frac{1}{3}}N^{\frac{2}{3}}q^{\frac{1}{9}+\frac{\delta}{2}}d^{\frac{2}{3}}.$$

We denote by $I(\lambda)$ the number of solutions of the congruence

$$(nd + \eta k)y^{-1} \equiv \lambda \pmod{q}, \quad M - N < n \leq M, \quad 1 \leq y \leq Y, \quad (y,q) = 1,$$

and obtain

$$|S| \leq (Y_q Z)^{-1} W + 2^{-\frac{1}{3}} N^{\frac{2}{3}} q^{\frac{1}{9}+\frac{\delta}{2}} d^{\frac{2}{3}}, \tag{13}$$

$$W = \sum_{\lambda=0}^{q-1} I(\lambda) \left| \sum_{1 \leq z \leq Z} \chi_q(\lambda + zd) \right|.$$

Cubing both sides of the equality, using Hölder's inequality and using the inequality

$$\sum_{\lambda=1}^{q} I(\lambda) \leq N Y_q,$$

we obtain

$$W^3 \leq \left(N Y_q\right)^2 \sum_{\lambda=0}^{q-1} I(\lambda) \left| \sum_{1 \leq z \leq Z} \chi_q(\lambda + zd) \right|^3.$$

By squaring both sides of the last inequality and applying Cauchy-Schwarz inequality, we have

$$W^6 \leq \left(N Y_q\right)^4 K V, \qquad K = \sum_{\lambda=0}^{q-1} I^2(\lambda), \qquad V = \sum_{\lambda=0}^{q-1} \left| \sum_{1 \leq z \leq Z} \chi_q(\lambda + zd) \right|^6.$$

Applying Lemma 4, we obtain

$$V = \sum_{\lambda=0}^{q-1} \sum_{z_1,\dots,z_6=1}^{Z} \chi\left(\frac{(\lambda + z_1 d)(\lambda + z_2 d)(\lambda + z_3 d)}{(\lambda + z_4 d)(\lambda + z_5 d)(\lambda + z_6 d)}\right)$$

$$\leq \sum_{z_1,\dots,z_6=1}^{Z} \left| \sum_{\lambda=0}^{q-1} \chi\left(\frac{(\lambda + z_1 d)(\lambda + z_2 d)(\lambda + z_3 d)}{(\lambda + z_4 d)(\lambda + z_5 d)(\lambda + z_6 d)}\right) \right|$$

$$\leq \sum_{z_1,\dots,z_6=1}^{Zd} \left| \sum_{\lambda=0}^{q-1} \chi\left(\frac{(\lambda + z_1)(\lambda + z_2)(\lambda + z_3)}{(\lambda + z_4)(\lambda + z_5)(\lambda + z_6)}\right) \right| \leq Z^3 d^3 q^{1+\delta}.$$

Therefore

$$W^6 \leq \left(N Y_q\right)^4 Z^3 d^3 q^{1+\delta} \cdot K. \tag{14}$$

The sum K equals the number of solutions of the congruence

$$(nd + \eta k)y^{-1} \equiv (n_1 d + \eta k)y_1^{-1} \pmod{q},$$

$$M < n, n_1 \le M + N, \quad 1 \le y, y_1 \le Y, \quad (y, q) = 1, \quad (y_1, q) = 1,$$

or the congruence

$$(nd + \eta k)y \equiv (n_1 d + \eta k)y_1 \pmod{q},$$

$$M < n, n_1 \le M + N, \quad 1 \le y, y_1 \le Y, \quad (y, q) = 1, \quad (y_1, q) = 1.$$

All conditions of Lemma 8 are satisfied for this congruence:

$$2NY = 2N \left[0.5 Nq^{-\frac{1}{6}} d\right] \le N^2 q^{-\frac{1}{6}} d < \left(q^{\frac{7}{12}} d^{-\frac{1}{2}}\right)^2 q^{-\frac{1}{6}} d = q,$$

$$\frac{Y}{d} = \frac{\left[0.5 Nq^{-\frac{1}{6}} d\right]}{d} > \frac{\left[0.5 q^{\frac{1}{6}} d\right]}{d} > 0.3 q^{\frac{1}{6}} > 1.$$

According to this Lemma, we have

$$K \le NY + \frac{2Y^2}{d} + \frac{2Y^2}{d}\rho(qd^{-1}, Y) + \frac{2(NY)^{1+\delta}}{d}$$

$$= \frac{2(NY)^{1+\delta}}{d} \left(1 + \frac{d}{2(NY)^\delta} + \frac{Y}{N(NY)^\delta}\left(\rho(qd^{-1}, Y) + 1\right)\right).$$

Taking into account that

$$d \le \exp(\sqrt{2\mathscr{L}}) \le \exp(2\sqrt{\mathscr{L}_q}) \le q^{\frac{\delta}{4}}, \quad \rho(qd^{-1}, Y) + 1 \le \tau(q) \le q^\delta,$$

$$(NY)^\delta \ge (0.1 \, N^2 q^{-\frac{1}{6}} d)^\delta > (0.1)^\delta q^{\frac{\delta}{2}},$$

we have

$$K \le \frac{2(NY)^{1+\delta}}{d} \left(1 + \frac{10^\delta q^{\frac{\delta}{4}}}{2q^{\frac{\delta}{2}}} + \frac{10 \, q^\delta}{q^{\frac{1}{6}+\frac{\delta}{2}}}\right) \le \frac{3(NY)^{1+\delta}}{d}.$$

Inserting this estimate into (14), and the right-hand side of obtained estimate into (13), we obtain

$$W^6 \le \left(NY_q\right)^4 Z^3 d^3 q^{1+\delta} \cdot K \le 3q N^5 Y Y_q^4 Z^3 d^2 (qNY)^\delta,$$

$$|S| \le \frac{3^{\frac{1}{6}} q^{\frac{1}{6}} N^{\frac{5}{6}} Y^{\frac{1}{6}} d^{\frac{1}{3}} (qNY)^{\frac{\delta}{6}}}{Y_q^{\frac{1}{3}} Z^{\frac{1}{2}}} + 2^{-\frac{1}{3}} N^{\frac{2}{3}} q^{\frac{1}{9}+\frac{\delta}{2}} d^{\frac{2}{3}}. \qquad (15)$$

Further, using the Lemma 5 and well-known inequalities

$$\omega(q) \leq \frac{c_\omega \mathscr{L}_q}{\ln \mathscr{L}_q}, \qquad \frac{\varphi(q)}{2q} \leq \frac{c_\varphi}{\ln \mathscr{L}_q},$$

where c_ω and c_φ are absolute constants, we find

$$\left| Y_q - \frac{\varphi(q)}{q} Y \right| \leq q^{\frac{c_\omega \ln 2}{\ln \mathscr{L}_q}} < \frac{\varphi(q)}{2q} q^{\frac{1}{7}} < \frac{\varphi(q)}{2q} \left[0.5 N q^{-\frac{1}{6}} \right] = \frac{\varphi(q)}{2q} Y,$$

i.e.,

$$Y_q > \frac{\varphi(q)}{2q} Y \geq \frac{c_\varphi Y}{\ln \mathscr{L}_q}.$$

Inserting the last inequality in (15), we express the parameter Y_q in terms of Y. We have

$$|S| \leq \frac{3^{\frac{1}{6}}}{c_\varphi^{\frac{1}{6}}} \cdot \frac{q^{\frac{1}{6}} N^{\frac{5}{6}} d^{\frac{1}{3}} (qNY)^{\frac{\delta}{6}} (\ln \mathscr{L}_q)^{\frac{1}{6}}}{Y^{\frac{1}{6}} Z^{\frac{1}{2}}} + 2^{-\frac{1}{3}} N^{\frac{2}{3}} q^{\frac{1}{9}+\frac{\delta}{2}} d^{\frac{2}{3}}.$$

Taking into account that $Y = \left[0.5 N q^{-\frac{1}{6}} d \right]$, $Z = \left[0.5 q^{\frac{1}{6}} d^{-1} \right]$ and $N < q^{\frac{7}{12}} d^{-\frac{1}{2}}$, we find

$$|S| \leq \frac{3^{\frac{1}{6}}}{c_\varphi^{\frac{1}{6}}} \cdot \frac{q^{\frac{1}{6}} N^{\frac{5}{6}} d^{\frac{1}{3}} \left(0.5 q^{\frac{5}{6}} N^2 d \right)^{\frac{\delta}{6}} (\ln \mathscr{L}_q)^{\frac{1}{6}}}{(0.25 N q^{-\frac{1}{6}} d)^{\frac{1}{6}} (0.25 q^{\frac{1}{6}} d^{-1})^{\frac{1}{2}}} + 2^{-\frac{1}{3}} N^{\frac{2}{3}} q^{\frac{1}{9}+\frac{\delta}{2}} d^{\frac{2}{3}}$$

$$= \frac{2^{\frac{2}{3}-\frac{\delta}{6}} 3^{\frac{1}{6}}}{c_\varphi^{\frac{1}{6}}} \cdot q^{\frac{1}{9}} N^{\frac{2}{3}} d^{\frac{2}{3}} \left(q^{\frac{1}{6}} N^2 d \right)^{\frac{\delta}{6}} (\ln \mathscr{L}_q)^{\frac{1}{6}} + 2^{-\frac{1}{3}} N^{\frac{2}{3}} q^{\frac{1}{9}+\frac{\delta}{2}} d^{\frac{2}{3}}$$

$$\leq \frac{2^{\frac{2}{3}-\frac{\delta}{6}} 3^{\frac{1}{6}}}{c_\varphi^{\frac{1}{6}}} \cdot N^{\frac{2}{3}} q^{\frac{1}{9}} d^{\frac{2}{3}} q^{\frac{\delta}{3}} (\ln \mathscr{L}_q)^{\frac{1}{6}} + 2^{-\frac{1}{3}} N^{\frac{2}{3}} q^{\frac{1}{9}+\frac{\delta}{2}} d^{\frac{2}{3}}$$

$$= \left(2^{\frac{2}{3}-\frac{\delta}{6}} 3^{\frac{1}{6}} c_\varphi^{-\frac{1}{6}} q^{-\frac{\delta}{6}} (\ln \mathscr{L}_q)^{\frac{1}{6}} + 2^{-\frac{1}{3}} \right) N^{\frac{2}{3}} q^{\frac{1}{9}+\frac{\delta}{2}} d^{\frac{2}{3}} \leq N^{\frac{2}{3}} q^{\frac{1}{9}+\frac{\delta}{2}} d^{\frac{2}{3}}.$$

This completes the proof of Lemma.

Lemma 10 *Let* $(\eta v, q) = 1$, $y \geq q^{\frac{1}{3}+\frac{8}{3}\delta}$, $y \leq q$, $D^{\frac{1}{2}} \leq q \leq D$ *and* $v \leq \exp(\sqrt{2\mathscr{L}})$, *then*

$$S_y(u, \eta, v) = \sum_{\substack{u-y < n \leq u \\ (n,q)=1, n \equiv \eta \ (\mathrm{mod} \ v)}} \chi_q(n - \eta) \ll \frac{y}{v} \exp\left(-0.7\sqrt{\mathscr{L}} \right).$$

Proof We have an identity

$$S_y(u, \eta, v) = \sum_{d|q} \mu(d) \sum_{\substack{u-y<nd\leq u \\ nd\equiv\eta \ (\mathrm{mod}\ v)}} \chi_q(nd - \eta).$$

Determining the number d_v^{-1} from the congruence $dd_v^{-1} \equiv 1(\mathrm{mod}\ v)$, we can rewrite the congruence $nd \equiv \eta(\mathrm{mod}\ v)$ as $n \equiv \eta d_v^{-1}(\mathrm{mod}\ v)$. Further, representing the variable of the summation n as $n = \eta d_v^{-1} + mv$, we have

$$S_y(u, \eta, v) = \sum_{d|q} \mu(d) \sum_{u-y<(\eta d_v^{-1}+mv)d\leq u} \chi_q\left(\left(\eta d_v^{-1} + mv\right)d - \eta\right)$$

$$= \sum_{d|q} \mu(d) \sum_{\frac{u-y-\eta dd_v^{-1}}{dv}<m\leq\frac{u-\eta dd_v^{-1}}{dv}} \chi_q\left(mvd + \eta\left(dd_v^{-1} - 1\right)\right).$$

Representing the congruence $dd_v^{-1} \equiv 1(\mathrm{mod}\ v)$ in the form $dd_v^{-1} - 1 = vk$, where $k = k(d, v)$ is uniquely defined for each pair of d and v, we have

$$S_y(u, \eta, v) = \chi_q(v) \sum_{d|q} \mu(d) S_y(u, \eta, v, d) \ll \sum_{d|q} \mu^2(d)|S_y(u, \eta, v, d)|,$$

$$(16)$$

$$S_y(u, \eta, v, d) = \sum_{u_1-y_1<m\leq u_1} \chi_q(md + \eta k), \quad u_1 = \frac{u-\eta}{dv} - \frac{\eta k}{d}, \quad y_1 = \frac{y}{dv}.$$

Part of the sum in the right side of (16), corresponding to the terms satisfying $d \leq \exp(\sqrt{2\mathscr{L}})$, will be denoted by $S_y'(u, \eta, v)$. The remaining terms are estimated using the trivial estimate for the sum $S_y(u, \eta, v, d)$ and the Lemma 7:

$$\sum_{\substack{d|q \\ d>\exp(\sqrt{2\mathscr{L}})}} \mu^2(d)|S_y(u, \eta, v, d)| \leq \sum_{\substack{d|q \\ \exp(\sqrt{2\mathscr{L}})<d}} \mu^2(d)\left(\frac{y}{vd} + 1\right)$$

$$\leq \frac{y}{v} \sum_{\substack{d|q \\ \exp(\sqrt{2\mathscr{L}})<d}} \frac{\mu^2(d)}{d} + \tau(q) \ll \frac{y}{v}\exp\left(-0.7\sqrt{\mathscr{L}}\right).$$

Therefore,

$$S_y(u, \eta, v) \ll \left|S_y'(u, \eta, v)\right| + \frac{y}{v}\exp\left(-0.7\sqrt{\mathscr{L}}\right). \quad (17)$$

To estimate $|S'_y(u, \eta, v)|$, we consider the following cases:

1. $y > \sqrt{q} \exp\left(\frac{c_\omega \mathscr{L}_q}{\ln \mathscr{L}_q}\right)$;
2. $q^{\frac{1}{3}+\frac{8}{5}\delta} \le y \le \sqrt{q} \exp\left(\frac{c_\omega \mathscr{L}_q}{\ln \mathscr{L}_q}\right)$.

1. To $S_y(u, \eta, v, d)$ we apply a formula that establishes a relation between the values of primitive characters and the values of Gauss sums (see [17, Ch. VIII, §1, Lemma 3]), which yields

$$
S_y(u, \eta, v, d) = \sum_{a=1}^{q} \sum_{\substack{u_1 - y_1 < m \le u_1, \\ a \equiv md + \eta k \pmod q}} \chi_q(md + \eta k)
$$

$$
= \sum_{a=1}^{q} \chi_q(a) \sum_{u_1 - y_1 < m \le u_1} \frac{1}{q} \sum_{t=0}^{q-1} e\left(\frac{(a - md - \eta k)t}{q}\right)
$$

$$
= \frac{\tau(\chi_q)}{q} \sum_{t=0}^{q-1} \bar{\chi}_q(t) e\left(\frac{-\eta kt}{q}\right) \sum_{u_1 - y_1 < m \le u_1} e\left(\frac{-mdt}{q}\right).
$$

It follows from the conditions $v \le \exp(\sqrt{2\mathscr{L}})$, $d \le \exp(\sqrt{2\mathscr{L}})$ and the hypothesis of the present case that the last sum over m is nonempty. Using the equality

$$
\sum_{m_1 \le m \le m_2} e\left(-\frac{mdt}{q}\right) = \frac{\sin \frac{\pi t d(m_2 - m_1 + 1)}{q}}{\sin \frac{\pi t d}{q}} e\left(-\frac{td(m_1 + m_2)}{2q}\right),
$$

with integers m_1 and m_2, passing to estimates, and taking into account that $|\tau(\chi)| = \sqrt{q}$, we find

$$
|S_y(u, \eta, v, d)| \le \frac{1}{\sqrt{q}} \sum_{\substack{t=0 \\ (t,q)=1}}^{q-1} \left|\sin \frac{\pi t}{q/d}\right|^{-1}
$$

$$
= \frac{1}{\sqrt{q}} \sum_{t_1=0}^{d-1} \sum_{\substack{t_2=0 \\ (q/dt_1 + t_2, q)=1}}^{q/d-1} \left|\sin \frac{\pi t_2}{q/d}\right|^{-1} \le \frac{d}{\sqrt{q}} \sum_{t=1}^{q/d-1} \left|\sin \frac{\pi t}{q/d}\right|^{-1}.
$$

If q/d is an odd number, then

$$
|S_y(u, \eta, v, d)| \le \frac{2d}{\sqrt{q}} \sum_{t=1}^{\frac{q/d-1}{2}} \frac{1}{\left|\sin \frac{\pi t}{q/d}\right|} \le \frac{2d}{\sqrt{q}} \sum_{t=1}^{\frac{q/d-1}{2}} \frac{q/d}{2t} = \sqrt{q} \sum_{t=1}^{\frac{q/d-1}{2}} \frac{1}{t},
$$

since $\sin \pi \alpha \geq 2\alpha$ for $0 \leq \alpha \leq 1/2$. Using the inequality $\frac{1}{t} \leq \ln \frac{2t+1}{2t-1}$, we then find

$$|S_y(u, \eta, v, d)| \leq \sqrt{q} \sum_{t=1}^{\frac{q/d-1}{2}} (\ln(2t+1) - \ln(2t-1)) = \sqrt{q} \ln q/d \leq \sqrt{q}\mathscr{L}_q.$$

If q/d is an even number, then

$$|S_y(u, \eta, v, d)| \leq \frac{2d}{\sqrt{q}} \sum_{t=1}^{\frac{q/d}{2}-1} \frac{1}{\left|\sin \frac{\pi t}{q/d}\right|} + \frac{d}{\sqrt{q}} \leq \frac{2d}{\sqrt{q}} \sum_{t=1}^{\frac{q/d}{2}-1} \frac{q/d}{2t} + \frac{d}{\sqrt{q}} \ll \sqrt{q}\mathscr{L}_q.$$

From this and from the definition of $S'_y(u, \eta, v)$ we have

$$S'_y(u, \eta, v) \ll \sum_{\substack{d|q \\ d \leq \exp(\sqrt{2\mathscr{L}})}} \mu^2(d)\sqrt{q}\mathscr{L}_q \leq \sqrt{q}\mathscr{L}_q \sum_{d|q} \mu^2(d) = 2^{\omega(q)} \sqrt{q}\mathscr{L}_q.$$

Using the inequality (2) and relations $\mathscr{L} \leq 2\mathscr{L}_q$, $v \leq \exp(\sqrt{2\mathscr{L}}) \leq \exp(2\sqrt{\mathscr{L}_q})$, we find

$$S'_y(u, \eta, v) \ll \frac{\sqrt{q}\,v \exp\left(\omega(q)\ln 2 + \ln \mathscr{L}_q + 0.7\sqrt{\mathscr{L}}\right)}{y} \cdot \frac{y}{v} \exp\left(-0.7\sqrt{\mathscr{L}}\right)$$

$$\leq \frac{\sqrt{q} \cdot \exp\left(c_\omega \ln 2 \frac{\mathscr{L}_q}{\ln \mathscr{L}_q} + 3\sqrt{\mathscr{L}_q} + \ln \mathscr{L}_q\right)}{y} \cdot \frac{y}{v} \exp\left(-0.7\sqrt{\mathscr{L}}\right)$$

$$\leq \frac{\sqrt{q} \cdot \exp\left(\frac{c_\omega \mathscr{L}_q}{\ln \mathscr{L}_q}\right)}{y} \cdot \frac{y}{v} \exp\left(-0.7\sqrt{\mathscr{L}}\right) \ll \frac{y}{v} \exp\left(-0.7\sqrt{\mathscr{L}}\right).$$

2. If $(d, k) > 1$, then $S_y(u, \eta, v, d) = 0$; therefore, without loss of generality, we will assume that $(d, k) = 1$ in $S_y(u, \eta, v, d)$. Applying Lemma 9, we obtain

$$|S_y(u, \eta, v, d)| \leq \left(\frac{y}{dv}\right)^{\frac{2}{3}} q^{\frac{1}{9}+\frac{\delta}{2}}d^{\frac{2}{3}} \leq \left(\frac{y}{v}\right)^{\frac{2}{3}} q^{\frac{1}{9}+\frac{\delta}{2}}.$$

From this and from the definition of $S'_y(u, \eta, v)$ we have

$$\left|S'_y(u, \eta, v)\right| \ll \left(\frac{y}{v}\right)^{\frac{2}{3}} q^{\frac{1}{9}+\frac{\delta}{2}} \sum_{\substack{d|q \\ d \leq \exp(\sqrt{2\mathscr{L}})}} \mu^2(d) \ll \left(\frac{y}{v}\right)^{\frac{2}{3}} q^{\frac{1}{9}+\frac{\delta}{2}}2^{\omega(q)}.$$

Using the inequality (2) and relation $\mathcal{L} \leq 2\mathcal{L}_q$, we find

$$\left| S_y'(u, \eta, v) \right| \ll \left(\frac{y}{v} \right)^{\frac{2}{3}} q^{\frac{1}{9} + \frac{\delta}{2}} \exp\left(c_\omega \ln 2 \frac{\mathcal{L}_q}{\ln \mathcal{L}_q} \right)$$

$$= \frac{y}{v} \exp\left(-0.7\sqrt{\mathcal{L}}\right) \cdot \left(\frac{q^{\frac{1}{3} + \frac{3\delta}{2}} v \cdot \exp\left(3c_\omega \ln 2 \frac{\mathcal{L}_q}{\ln \mathcal{L}_q} + 2.1\sqrt{\mathcal{L}}\right)}{y} \right)^{\frac{1}{3}}$$

$$< \frac{y}{v} \exp\left(-0.7\sqrt{\mathcal{L}}\right) \cdot \left(\frac{q^{\frac{1}{3} + \frac{3\delta}{2}} \cdot \exp\left(3c_\omega \ln 2 \frac{\mathcal{L}_q}{\ln \mathcal{L}_q} + 4\sqrt{\mathcal{L}_q}\right)}{y} \right)^{\frac{1}{3}}$$

$$< \frac{y}{v} \exp\left(-0.7\sqrt{\mathcal{L}}\right) \cdot \left(\frac{q^{\frac{1}{3} + \frac{8\delta}{5}}}{y} \right)^{\frac{1}{3}} \ll \frac{y}{v} \exp\left(-0.7\sqrt{\mathcal{L}}\right).$$

Inserting our estimate of $|S_y'(u, \eta, v)|$ in (17) completes the proof of the Lemma.

4 Estimates of Double Character Sums

Lemma 11 *Let M, N, U be integers, $N \leq U < 2N$, a_m and b_n are such integer-valued functions that*

$$\sum_{M < m \leq 2M} |a_m|^\alpha \ll M\mathcal{L}^{c_\alpha}, \quad \alpha = 1, 2; \qquad |b_n| \ll B.$$

Then the following estimate holds:

$$W = \sum_{M < m \leq 2M} a_m \sum_{\substack{U < n \leq \min(xm^{-1}, 2N) \\ (mn,q)=1, \, mn \equiv l(\text{mod } v)}} b_n \chi_q(mn - l)$$

$$\ll B\left(M^{\frac{3}{4}} N^{\frac{1}{2}} q^{\frac{1}{4}} + M^{\frac{3}{4}} Nq^{\frac{1}{8} + \frac{\delta}{4}} \right) \mathcal{L}^{\frac{2c_1 + c_2}{4} + 1}.$$

Proof It follows from the condition $(l, v) = 1$ that $(mn, v) = 1$. Denoting the inner sum in W by $\mathcal{B}(m)$ and representing the congruence $mn \equiv l(\text{mod } v)$ in the form $n \equiv lm_v^{-1}(\text{mod } v)$, we shall transform the inner sum so that it does not depend on m. We have the identity

$$\mathcal{B}(m) = \sum_{\substack{N < n \leq 2N \\ (n,q)=1}} b_n \chi_q(mn - l) \sum_{U < r \leq \min(xm^{-1}, 2N)}$$

$$\times \frac{1}{q} \sum_{k=0}^{q-1} e\left(\frac{k(n-r)}{q} \right) \frac{1}{v} \sum_{j=0}^{v-1} e\left(\frac{j(n - lm_v^{-1})}{v} \right)$$

$$= \frac{1}{qv}\sum_{j=0}^{v-1}e\left(-\frac{lm_v^{-1}j}{v}\right)\sum_{k=0}^{q-1}\mathscr{B}(kv+jq,m)\sum_{U<r\leq\min(xm^{-1},2N)}e\left(-\frac{kr}{q}\right),$$

$$\mathscr{B}(kv+jq,m)=\sum_{\substack{N<n\leq 2N\\(n,q)=1}}b_n\chi_q(mn-l)e\left(\frac{(kv+jq)n}{qv}\right).$$

Denoting $N'=\min([xm^{-1}],2N)$, extracting the term with $k=0$ and summing over r, we obtain

$$|\mathscr{B}(m)|\leq\frac{1}{qv}\sum_{j=0}^{v-1}e\left(-\frac{lm_v^{-1}j}{v}\right)\left((N'-U)|\mathscr{B}(jq,m)|\right.$$

$$\left.+\sum_{k=1}^{q-1}|\mathscr{B}(kv+jq,m)|\frac{\sin\frac{\pi k(N'-U)}{q}}{\sin\frac{\pi k}{q}}e\left(-\frac{k(N'+1+U)}{2q}\right)\right).$$

Passing to the inequalities, we find

$$|\mathscr{B}(m)|\leq\frac{1}{v}\sum_{j=0}^{v-1}\left(\frac{N'-N}{q}|\mathscr{B}(jq,m)|+\frac{1}{q}\sum_{k\leq[q/2]}\frac{|\mathscr{B}(kv+jq,m)|}{\left|\sin\frac{\pi k}{q}\right|}\right.$$

$$\left.+\frac{1}{q}\sum_{q/2<k\leq q-1}\frac{|\mathscr{B}(kv+jq,m)|}{\left|\sin\frac{\pi(q-k)}{q}\right|}\right).$$

Next, using the condition $N'-N<q$, inequalities $\sin\pi\alpha\geq\alpha$, $0\leq\alpha\leq 0.5$ and $\frac{1}{k}\leq\frac{2}{k+1}$, where $k\geq 1$ is an integer, we have

$$|\mathscr{B}(m)|$$

$$\leq\frac{1}{v}\sum_{j=0}^{v-1}\left(\mathscr{B}(jq,m)|+\sum_{k\leq[q/2]}\frac{|\mathscr{B}(kv+jq,m)|}{k}+\sum_{q/2<k\leq q-1}\frac{|\mathscr{B}(kv+jq,m)|}{q-k}\right)$$

$$\leq\frac{2}{v}\sum_{j=0}^{v-1}\sum_{k=0}^{q-1}\left(\frac{1}{k+1}+\frac{1}{q-k}\right)|\mathscr{B}(kv+jq,m)|$$

$$\ll\mathscr{L}\max_{0\leq j<v}\max_{0\leq k<q}|\mathscr{B}(kv+jq,m)|.$$

From this and from the definition of W, we have

$$|W|\ll\mathscr{L}\max_{0\leq j<v}\max_{0\leq k<q}W(j,k),\quad W(j,k)=\sum_{\substack{M<m\leq 2M\\(m,q)=1}}|a_m||\mathscr{B}(kv+jq,m)|.$$

$$(18)$$

Let us estimate $W(j, k)$. Squaring both sides of the last equality and applying Hölder's inequality, we have

$$
W^2(j, k) \leq \sum_{\substack{M < m \leq 2M \\ (m,q)=1}} |a_m| \sum_{\substack{M < m \leq 2M \\ (m,q)=1}} |a_m| |\mathscr{B}(kd + jq, m)|^2
$$

$$
\ll M \mathscr{L}^{c_1} \sum_{\substack{M < m \leq 2M \\ (m,q)=1}} |a_m| |\mathscr{B}(kd + jq, m)|^2.
$$

Now, squaring both sides of the last inequality and applying Cauchy-Schwarz inequality, we find

$$
W^4(j, k) \ll M^2 \mathscr{L}^{2c_1} \sum_{\substack{M < m \leq 2M \\ (m,q)=1}} |a_m|^2 \sum_{\substack{M < m \leq 2M \\ (m,q)=1}} |\mathscr{B}(kd + jq, m)|^4
$$

$$
\ll M^3 \mathscr{L}^{2c_1 + c_2} \sum_{\substack{m=0 \\ (m,q)=1}}^{q-1} \left| \sum_{\substack{N < n \leq 2N \\ (n,q)=1}} b_n \chi_q(n - lm_q^{-1}) e\left(\frac{(kd + jq)n}{qd} \right) \right|^4
$$

$$
= M^3 \mathscr{L}^{2c_1 + c_2} \sum_{\substack{\lambda=0 \\ (\lambda,q)=1}}^{q-1} \left| \sum_{\substack{N < n \leq 2N \\ (n,q)=1}} b_n \chi_q(n + \lambda) e\left(\frac{(kd + jq)n}{qd} \right) \right|^4
$$

$$
\ll M^3 \mathscr{L}^{2c_1 + c_2} \sum_{\substack{N' < n_{1,2},n_3,n_4 \leq 2N \\ (n_{1,2},n_3,n_4),q)=1}} |b_{n_1} b_{n_2} b_{n_3} b_{n_4}| \left| \sum_{\lambda=0}^{q-1} \chi\left(\frac{(\lambda + n_1)(\lambda + n_2)}{(\lambda + n_3)(\lambda + n_4)} \right) \right|.
$$

Next, using the estimate $|b_n| \ll B$, and the Lemma 3, we obtain

$$
W^4(j, k) \ll M^3 \mathscr{L}^{2c_1 + c_2} B^4 \sum_{1 \leq n_1,\dots n_4 \leq 2N} \left| \sum_{\lambda=0}^{q-1} \chi\left(\frac{(\lambda + n_1)(\lambda + n_2)}{(\lambda + n_3)(\lambda + n_4)} \right) \right|
$$

$$
\ll M^3 \mathscr{L}^{2c_1 + c_2} B^4 \left(N^2 q + N^4 q^{\frac{1}{2}+\delta} \right) \ll B^4 \left(M^3 N^2 q + M^3 N^4 q^{\frac{1}{2}+\delta} \right) \mathscr{L}^{2c_1 + c_2}.
$$

The Lemma now follows from this estimate and (18).

Corollary 1 Let M, N, U be integers, $N \leq U < 2N$, $q^{\frac{1}{4}-\theta} \leq N \leq q^{\frac{1}{4}+\theta}$, $D^{\frac{1}{2}} \leq q \leq D$, $v \leq \exp(\sqrt{2\mathscr{L}})$, a_m and b_n are such integer-valued functions that $|a_m| \leq \tau_5(m)$, $|b_n| \leq 1$. Then the following estimate holds for $x \geq q^{\frac{3}{4}+\theta+1.1\delta}$:

$$
W = \sum_{M < m \leq 2M} a_m \sum_{\substack{U < n \leq \min(xm^{-1}, 2N) \\ (mn,q)=1, \, mn \equiv l \,(\mathrm{mod}\, v)}} b_n \chi_q(mn - l) \ll \frac{x}{v} \exp\left(-0.7\sqrt{\mathscr{L}} \right).
$$

Proof Taking into account that $\ln M \ll \mathscr{L}$, we have from the Lemma 6

$$\sum_{M < m \leq 2M} \tau_5(m) \ll M\mathscr{L}^4, \qquad \sum_{M < m \leq 2M} \tau_5^2(m) \ll M\mathscr{L}^{24}.$$

Applying the Lemma 11 for $c_1 = 4$, $c_2 = 24$ and using conditions $MN \leq x$, $q^{\frac{1}{4}-\theta} \leq N \leq q^{\frac{1}{4}+\theta}$ and $x \geq q^{\frac{3}{4}+\theta+1.1\delta}$, we find that

$$W \ll \left(M^{\frac{3}{4}}N^{\frac{1}{2}}q^{\frac{1}{4}} + M^{\frac{3}{4}}Nq^{\frac{1}{8}}\right)\mathscr{L}^9\delta^{\frac{5}{4}} \leq x^{\frac{3}{4}}\left(N^{-\frac{1}{4}}q^{\frac{1}{4}} + N^{\frac{1}{4}}q^{\frac{1}{8}}\right)\mathscr{L}^9 q^{\frac{\delta}{4}}$$

$$\ll x\left(\frac{qN^{-1}}{x} + \frac{Nq^{\frac{1}{2}}}{x}\right)^{\frac{1}{4}}\mathscr{L}^9 q^{\frac{\delta}{4}} \ll x\left(\frac{q^{\frac{3}{4}+\theta}}{x}\right)^{\frac{1}{4}}\mathscr{L}^9 q^{\frac{\delta}{4}} \ll x\mathscr{L}^9 q^{-\frac{\delta}{40}}.$$

Applying relations $D^{\frac{1}{2}} \leq q \leq D$ and $v \leq \exp(\sqrt{2\mathscr{L}})$, we obtain

$$W \ll \frac{x}{v}\mathscr{L}^9 \exp(\sqrt{2\mathscr{L}})\, D^{-\frac{\delta}{80}} \ll \frac{x}{v}\exp\left(-0.7\sqrt{\mathscr{L}}\right).$$

Lemma 12 *Let* M, N, U *be integers,* $N \leq U < 2N \leq q^{\frac{1}{6}}$, a_m *and* b_n *are such integer-valued functions that*

$$\sum_{M < m \leq 2M} |a_m|^\alpha \ll M\mathscr{L}^{c_\alpha}, \qquad \alpha = 1, 2; \qquad |b_n| \ll B.$$

Then the following estimate holds:

$$W = \sum_{M < m \leq 2M} a_m \sum_{\substack{U < n \leq \min(xm^{-1}, 2N) \\ (mn,q)=1,\, mn\equiv l\,(\mathrm{mod}\, v)}} b_n \chi_q(mn - l) \ll BM^{\frac{5}{6}}N^{\frac{1}{2}}q^{\frac{1}{6}+\frac{1}{6}\delta}\mathscr{L}^{\frac{4c_1+c_2}{6}+1}.$$

Proof Without loss of generality, we can assume that $MN < x$. Repeating the argument used in the proof of the previous lemma, we find

$$|W| \ll \mathscr{L} \max_{0 \leq j < v} \max_{0 \leq k < q} W(j, k), \qquad W(j, k) = \sum_{\substack{M < m \leq 2M \\ (m,q)=1}} |a_m||\mathscr{B}(kv + jq, m)|.$$

$$(19)$$

$$\mathscr{B}(kv + jq, m) = \sum_{\substack{N < n \leq 2N \\ (n,q)=1}} b_n \chi_q(mn - l)e\left(\frac{(kv + jq)n}{qv}\right).$$

Let us estimate $W(j, k)$. Cubing both sides of the identity and applying Hölder's inequality, we have

$$W^3(j, k) \ll M^2 \mathcal{L}^{2c_1} \sum_{\substack{M < m \le 2M \\ (m,q)=1}} |a_m| |\mathcal{B}(kd + jq, m)|^3.$$

Now, squaring both sides of the last estimate and applying Cauchy-Schwarz inequality, we find

$$W^6(j, k) \ll M^4 \mathcal{L}^{4c_1} \sum_{\substack{M < m \le 2M \\ (m,q)=1}} |a_m|^2 \sum_{\substack{M < m \le 2M \\ (m,q)=1}} |\mathcal{B}(kd + jq, m)|^6$$

$$\ll M^5 \mathcal{L}^{4c_1+c_2} \sum_{\substack{m=0 \\ (m,q)=1}}^{q-1} \left| \sum_{\substack{N < n \le 2N \\ (n,q)=1}} b_n \chi_q(n - lm_q^{-1}) e\left(\frac{(kd + jq)n}{qd}\right) \right|^6$$

$$= M^5 \mathcal{L}^{4c_1+c_2} \sum_{\substack{\lambda=0 \\ (\lambda,q)=1}}^{q-1} \left| \sum_{\substack{N < n \le 2N \\ (n,q)=1}} b_n \chi_q(n + \lambda) e\left(\frac{(kd + jq)n}{qd}\right) \right|^6$$

$$\ll M^5 \mathcal{L}^{4c_1+c_2} \sum_{\substack{N' < n_1,\dots,n_6 \le 2N \\ (n_1,\dots,n_6),q)=1}} |b_{n_1} \dots b_{n_6}| \left| \sum_{\lambda=0}^{q-1} \chi\left(\frac{(\lambda+n_1)(\lambda+n_2)(\lambda+n_3)}{(\lambda+n_4)(\lambda+n_5)(\lambda+n_6)}\right) \right|.$$

Next, using the estimate $|b_n| \ll B$ and the Lemma 4, we find that

$$W^6(j, k) \ll B^6 M^5 N^3 q^{1+\delta} \mathcal{L}^{4c_1+c_2}.$$

The Lemma now follows from this estimate and (19).

Corollary 2 *Let M, N, U be integers, $N \le U < 2N$, $q^\theta \le N \le q^{\frac{1}{6}}$, $D^{\frac{1}{2}} \le q \le D$, $v \le \exp(\sqrt{2\mathcal{L}})$, a_m and b_n are such integer-valued functions that $|a_m| \le \tau_5(m)$, $|b_n| \le 1$. Then the following estimate holds for $x \ge q^{1-2\theta+1.1\delta}$:*

$$W = \sum_{M < m \le 2M} a_m \sum_{\substack{U < n \le \min(xm^{-1}, 2N) \\ (mn,q)=1,\, mn \equiv l \pmod{v}}} b_n \chi_q(mn - l) \ll \frac{x}{v} \exp\left(-0.7\sqrt{\mathcal{L}}\right).$$

Proof Taking into account that $\ln M \ll \mathcal{L}$, we have from the Lemma 6 that

$$\sum_{M < m \le 2M} \tau_5(m) \ll M\mathcal{L}^4, \qquad \sum_{M < m \le 2M} \tau_5^2(m) \ll M\mathcal{L}^{24}.$$

Applying the Lemma $x \geq q^{1-2\theta+1.1\delta}$ for $c_1 = 4$, $c_2 = 24$ and using conditions $MN \leq x$, $N \geq q^{\theta}$ and $x \geq q^{1-2\theta+1.1\delta}$, we find that

$$W \ll (MN)^{\frac{5}{6}} N^{-\frac{1}{3}} q^{\frac{1}{6}+\frac{1}{6}\delta} \mathscr{L}^{\frac{20}{3}} \leq x^{\frac{5}{6}} N^{-\frac{1}{3}} q^{\frac{1}{6}+\frac{1}{6}\delta} \mathscr{L}^{\frac{20}{3}} =$$

$$= x \left(\frac{N^{-2} q^{1+\delta}}{x} \right)^{\frac{1}{6}} \mathscr{L}^{\frac{20}{3}} \leq x \left(\frac{q^{1-2\theta+\delta}}{x} \right)^{\frac{1}{6}} \mathscr{L}^{\frac{20}{3}} \ll x q^{-\frac{\delta}{60}} \mathscr{L}^{\frac{20}{3}}.$$

Applying relations $D^{\frac{1}{2}} \leq q \leq D$ and $v \leq \exp(\sqrt{2\mathscr{L}})$, we obtain

$$W \ll \frac{x}{v} \mathscr{L}^{\frac{20}{3}} \exp(\sqrt{2\mathscr{L}}) D^{-\frac{\delta}{120}} \ll \frac{x}{v} \exp\left(-0.7\sqrt{\mathscr{L}}\right).$$

5 Proof of the Theorem 1

Without loss of generality, we shall assume that

$$x = D^{\frac{5}{6}+\varepsilon} \geq q^{\frac{5}{6}+\varepsilon}.$$

Having in mind that the contribution of terms satisfying $(n, q) > 1$ in the sum $T(\chi)$ has order of magnitude $\ll \mathscr{L}^2$ and using that χ_q is a primitive character generated by a nonprincipal character χ and q_1 is a product of primes, dividing D, but not q, we have

$$T(\chi) = \sum_{\substack{n \leq x \\ (n,q)=1}} \Lambda(n) \chi(n-l) + O(\mathscr{L}^2) = \sum_{\substack{n \leq x, (n,q)=1 \\ (n-l,q_1)=1}} \Lambda(n) \chi_q(n-l) + O(\mathscr{L}^2)$$

$$= \sum_{v|q_1} \mu(v) T(\chi_q, v) + O(\mathscr{L}^2), \qquad T(\chi_q, v) = \sum_{\substack{n \leq x, (n,q)=1 \\ n \equiv l \pmod{v}}} \Lambda(n) \chi_q(n-l).$$

Part of the sum $T(\chi)$ corresponding to terms satisfying the condition $\exp(\sqrt{2\mathscr{L}}) < v \leq x$ shall be denoted by $T_1(\chi)$. Let us estimate $T_1(\chi)$ using the trivial estimate for the sum $T(\chi_q, v)$ and the Lemma 7:

$$|T_1(\chi)| \ll \mathscr{L} \sum_{\substack{v|q_1 \\ v > \exp \sqrt{2\mathscr{L}}}} \mu^2(v) \left(\frac{x}{v} + 1 \right) \ll x\mathscr{L} \sum_{\substack{v|q_1 \\ v > \exp \sqrt{2\mathscr{L}}}} \frac{\mu^2(v)}{v} \ll x\mathscr{L} \exp\left(-0.7\sqrt{\mathscr{L}}\right).$$

Therefore

$$T(\chi) = \sum_{\substack{v|q_1 \\ v \leq \exp(\sqrt{2\mathscr{L}})}} \mu(v) T(\chi_q, v) + O\left(x\mathscr{L} \exp\left(-0.7\sqrt{\mathscr{L}}\right)\right). \qquad (20)$$

Let us estimate $T(\chi_q, v)$ for $v \le \exp\sqrt{2\mathscr{L}}$ and $(v, l) = (q, l) = (v, q) = 1$. The sum $T(\chi_q, v)$ is studied in detail in the Lemma 5 of [28] and we use the following estimate

$$|T(\chi_q, v)| \le 10x \ln^5 x \left(\sqrt{\frac{1}{qv^2} + \frac{q}{x}} + x^{-\frac{1}{6}} v^{-\frac{1}{2}} + x^{-\frac{1}{3}} q^{\frac{1}{6}} v^{-\frac{1}{3}} \right) \tau(q).$$

For $x > q^{1+1.2\varepsilon}$ and $v \le \exp\sqrt{2\mathscr{L}}$ the last estimate gives the following nontrivial estimate for the sum

$$|T(\chi_q, v)| \ll \frac{x}{v} \exp\left(-0.7\sqrt{\mathscr{L}}\right).$$

Thus, henceforth we shall assume that $x \le q^{1+1.2\varepsilon}$, i.e.

$$D^{\frac{5}{6}} \le q \le D. \tag{21}$$

Setting $u = x^{\frac{1}{3}}$, $r = 3$ in the Lemma 1 and

$$f(n) = \begin{cases} \chi_q(n - l), & \text{for } (n, q) = 1 \text{ and } n \equiv l \ (\mathrm{mod}\ v); \\ 0, & \text{otherwise}, \end{cases}$$

we find

$$T(\chi_q, v) = \sum_{k=1}^{3} (-1)^k C_3^k \tilde{T}_k(\chi_q, v), \tag{22}$$

$$\tilde{T}_k(\chi_q, v) = \sum_{m_1 \le u} \mu(m_1) \cdots \sum_{m_k \le u} \mu(m_k) \sum_{n_1} \cdots \sum_{\substack{n_k \\ m_1 \cdots m_k n_1 \cdots n_k \le x, \ (m_1 \cdots m_k n_1 \cdots n_k, q)=1, \ m_1 \cdots m_k n_1 \cdots n_k \equiv l \,(\mathrm{mod}\ v)}} \ln n_1 \chi_q(m_1 n_1 \cdots m_k n_k - l).$$

Let us divide in $T_k(\chi_q, v)$ the limits of each variable $m_1, \cdots, m_k, n_1, \cdots, n_k$ into not more than $\mathscr{L}$ intervals of the form $M_j < m_j \le 2M_j$, $N_j < n_j \le 2N_j$, $j = 1, 2, \cdots, k$. We obtain not more than $\mathscr{L}^{2k}$ sums of the form

$$\hat{T}_k(\chi_q, v)$$

$$= \sum_{M_1 < m_1 \le 2M_1} \mu(m_1) \cdots \sum_{M_k < m_k \le 2M_k} \mu(m_k) \sum_{N_1 < n_1 \le 2N_1} \cdots \sum_{\substack{N_k < n_k \le 2N_k \\ m_1 n_1 \cdots m_k n_k \le x, \ (m_1 n_1 \cdots m_k n_k, q)=1, \ m_1 n_1 \cdots m_k n_k \equiv l\,(\mathrm{mod}\ v)}} \chi_q(m_1 n_1 \cdots m_k n_k - l) \ln n_1$$

$$= \int_1^{2N_1} \sum_{M_1 < m_1 \le 2M_1} \mu(m_1) \cdots \sum_{M_k < m_k \le 2M_k} \mu(m_k) \sum_{\substack{\max(u, N_1) < n_1 \le 2N_1 \\ m_1 n_1 \cdots m_k n_k \le x, \ (m_1 n_1 \cdots m_k n_k, q)=1, \ m_1 n_1 \cdots m_k n_k \equiv l\,(\mathrm{mod}\ v)}}$$

$$\cdots \sum_{N_k < n_k \le 2N_k} \chi_q(m_1 n_1 \cdots m_k n_k - l) d \ln u.$$

Let us denote by $U_1 = \max(u, N_1)$ such number u for which the integrand takes on its maximum value. Then we have the following inequality $|\hat{T}_k(\chi_q, v)| \ll \mathcal{L}\,|T_k(\chi_q, v)|$, where

$$T_k(\chi_q, v) = \sum_{\substack{M_1 < m_1 \leq 2M_1 \\ m_1 n_1 \cdots m_k n_k \leq x,\ (m_1 n_1 \cdots m_k n_k, q) = 1,}} \mu(m_1) \cdots \sum_{\substack{M_k < m_k \leq 2M_k}} \mu(m_k) \sum_{\substack{U_1 < n_1 \leq 2N_1}} \cdots \sum_{\substack{U_k < n_k \leq 2N_k \\ m_1 n_1 \cdots m_k n_k \equiv l \pmod v}} \chi_q(m_1 n_1 \cdots m_k n_k - l),$$

$$x^{\frac{1}{3}} > M_1 \geq M_2 \geq \cdots \geq M_k, \quad N_1 \geq N_2 \geq \cdots \geq N_k, N_j \leq U_j < 2N_j. \tag{23}$$

From this and the estimates (22), (20), we obtian

$$|T(\chi)| \ll \sum_{\substack{v \mid q_1 \\ v \leq \exp(\sqrt{2\mathcal{L}})}} \mu^2(v) \sum_{k=1}^{3} \mathcal{L}^6 \max |T_k(\chi_q, v)| + x \exp\left(-0.6\sqrt{\mathcal{L}}\right). \tag{24}$$

Let us introduce the following notations:

$$\prod_{j=1}^{k} M_j N_j = Y, \qquad \prod_{j=1}^{k} M_j U_j = X, \qquad Y < X \leq x,$$

and further we shall assume that

$$Y \geq x \exp\left(-\sqrt{\mathcal{L}}\right), \tag{25}$$

since otherwise, estimating $T_k(\chi_q, v)$ trivially, we have

$$T_k(\chi_q, v) \ll \sum_{\substack{X < n \leq 2^k Y \\ n \equiv l \pmod v}} \tau_{2k}(n) \ll \frac{2^k Y}{v} \mathcal{L}^{2k-1}$$

$$\leq \frac{x}{v} 2^k \mathcal{L}^{2k-1} \exp\left(-\sqrt{\mathcal{L}}\right) \ll \frac{x}{v} \exp\left(-0.7\sqrt{\mathcal{L}}\right).$$

The sums $T_k(\chi_q, v)$, $k = 1, 2, 3$ are estimated in the same way. For example, we shall estimate the sum $T_3(\chi_q, v)$ and consider the following possible values for the parameter N_1:

1. $N_1 > q^{\frac{1}{3} + \frac{16}{27}\varepsilon}$;
2. $q^{\frac{1}{6}} < N_1 \leq q^{\frac{1}{3} + \frac{16}{27}\varepsilon}$;
3. $q^{\frac{1}{12}} < N_1 \leq q^{\frac{1}{6}}$,
4. $N_1 < q^{\frac{1}{12}}$.

In considering the cases 1, 2, we shall transform the sum $T_3(\chi_q, \nu)$ and rewrite it in the form

$$T_3(\chi_q, \nu) = \sum_{\substack{XU_1^{-1} < m \le 2^5 Y N_1^{-1}}} a_m \sum_{\substack{U_1 < n \le 2N_1, \, mn \le x \\ (mn,q)=1, \, mn \equiv l \,(\mathrm{mod}\, \nu)}} \chi_q(mn - l), \qquad |a_m| \le \tau_5(m).$$

Next, we shall divide the interval of summation $XU_1^{-1} < m \le 2^5 Y N_1^{-1}$ into smaller intervals of the form $M < m \le 2M$. We shall obtain not more than five sums of the form

$$T_3(\chi_q, \nu, M) = \sum_{M < m \le 2M} a_m \sum_{\substack{U_1 < n \le \min(xm^{-1}, 2N_1) \\ (mn,q)=1, \, mn \equiv l \,(\mathrm{mod}\, \nu)}} \chi_q(mn - l).$$

Case 1. $N_1 > q^{\frac{1}{3} + \frac{16}{27}\varepsilon}$. Determining the number m_q^{-1} from the congruence $mm_q^{-1} \equiv 1 \ (\mathrm{mod}\ q)$ and passing to the estimate, we find

$$|T_3(\chi_q, \nu, M)| \le \sum_{\substack{M < m \le 2M \\ (m,q)=1}} \tau_5(m) \left| \sum_{\substack{U_1 < n \le \min(xm^{-1}, 2N_1) \\ (n,q)=1, \, mn \equiv l \,(\mathrm{mod}\, \nu)}} \chi_q(mn - l) \right|.$$

Applying the Lemma 10 to the sum over n with

$$\delta = \frac{10}{27}\varepsilon, \quad \eta = lm_q^{-1}, \quad u = \min(xm^{-1}, 2N_1), \quad y = \min(xm^{-1}, 2N_1) - U_1 \le N_1,$$

and using the condition $N_1 > q^{\frac{1}{3} + \frac{16}{27}\varepsilon}$, we have

$$|T_3(\chi_q, \nu, M)| \ll M \mathscr{L}^4 \cdot \frac{N_1}{\nu} \exp\left(-0.7\sqrt{\mathscr{L}}\right)$$

$$\le \frac{2^5 Y}{\nu} \mathscr{L}^4 \exp\left(-0.7\sqrt{\mathscr{L}}\right) \ll \frac{x}{\nu} \mathscr{L}^4 \exp\left(-0.7\sqrt{\mathscr{L}}\right).$$

Case 2. $q^{\frac{1}{6}} < N_1 \le q^{\frac{1}{3} + \frac{16}{27}\varepsilon}$. We shall use the Corollary 1 of Lemma 11 with

$$U = U_1, \qquad N = N_1, \qquad \theta = \frac{1}{12} + \frac{16}{27}\varepsilon, \qquad \delta = \frac{10}{27}\varepsilon, \qquad b_n = 1.$$

Then, for $x \ge q^{\frac{3}{4} + \theta + 1.1\delta} = q^{\frac{5}{6} + \varepsilon}$ we have

$$|T_3(\chi_q, \nu, M)| \ll \frac{x}{\nu} \exp\left(-0.7\sqrt{\mathscr{L}}\right).$$

Case 3. $q^{\frac{1}{12}} < N_1 \leq q^{\frac{1}{6}}$.The conditions of the Corollary 2 are satisfied for the sum $T_3(\chi_q, \nu, M)$ for

$$U = U_1, \qquad N = N_1, \qquad \theta = \frac{1}{12}, \qquad b_n = 1, \qquad \delta = \frac{10}{27}\varepsilon.$$

Applying this Corollary for $x \geq q^{1-2\theta+1.1\delta} = q^{\frac{5}{6}+\frac{11}{27}\varepsilon}$, we obtain

$$|T_3(\chi_q, \nu, M)| \ll \frac{x}{\nu} \exp\left(-0.7\sqrt{\mathscr{L}}\right).$$

Case 4. $N_1 < q^{\frac{1}{12}}$. We shall transform the sum $T_3(\chi_q, \nu)$ by rewriting it in the form

$$T_3(\chi_q, \nu) = \sum_{XM_1^{-1} < m \leq 2^5 Y M_1^{-1}} a_m \sum_{\substack{M_1 < n \leq 2M_1, \, mn \leq x \\ (mn,q)=1, \, mn \equiv l \,(\mathrm{mod}\ \nu)}} \mu(n)\chi_q(mn - l), \quad |a_m| \leq \tau_5(m),$$

and divide the interval of summation $XM_1^{-1} < m \leq 2^5 Y M_1^{-1}$ into smaller intervals of the form $M < m \leq 2M$. We shall obtain not more than five sums of the form

$$T_3(\chi_q, \nu, M) = \sum_{M < m \leq 2M} a_m \sum_{\substack{M_1 < n \leq \min(xm^{-1}, 2M_1) \\ (mn,q)=1, \, mn \equiv l \,(\mathrm{mod}\ \nu)}} \mu(n)\chi_q(mn - l).$$

Using the relations (23), (25), conditions of the case 4 and the relations (21), we have

$$M_1 \geq (M_1 M_2 M_3)^{\frac{1}{3}} = \left(\frac{Y}{N_1 N_2 N_3}\right)^{\frac{1}{3}} \geq \frac{Y^{\frac{1}{3}}}{N_1} \geq \frac{\left(x \exp\left(-\sqrt{\mathscr{L}}\right)\right)^{\frac{1}{3}}}{N_1}$$

$$\geq \frac{q^{\frac{5}{18}+\frac{5}{18}\varepsilon} \exp\left(-0.4\sqrt{\mathscr{L}}\right)}{q^{\frac{1}{12}}} > q^{\frac{7}{36}},$$

$$M_1 \leq x^{\frac{1}{3}} = D^{\frac{5}{18}+\frac{\varepsilon}{3}} \leq q^{\frac{1}{3}+\frac{2}{5}\varepsilon}.$$

It follows that the conditions of the Corollary 1 are satisfied for the sum $T_3(\chi_q, \nu, M)$ for

$$U = M_1, \qquad N = M_1, \qquad \theta = \frac{1}{12} + \frac{2}{5}\varepsilon, \qquad \delta = \frac{10}{27}\varepsilon, \qquad b_n = \mu(n).$$

Applying this corollary for

$$x \geq q^{\frac{3}{4}+\theta+1.1\delta} = q^{\frac{5}{6}+\frac{109}{135}\varepsilon},$$

we obtain

$$|T_3(\chi_q, \nu, M)| \ll \frac{x}{\nu} \exp\left(-0.7\sqrt{\mathscr{L}}\right).$$

The Lemma now follows by inserting the estimates $T_k(\chi_q, \nu)$, $k = 1, 2, 3$ into (24).

References

1. D.A. Burgess, On character sums and L – series. Proc. Lond. Math. Soc. **12**(3), 193–206 (1962)
2. D.A. Burgess, The character sum estimate with $r = 3$. J. Lond. Math. Soc. **33**, 219–226 (1986)
3. J.B. Friedlander, K. Gong, I.E. Shparlinski, Character sums over shifted primes. Math. Notes **88**(3–4), 585–598 (2010)
4. D.R. Heath-Brown, Prime numbers in short intervals and a generalized Vaughan identity. Can. J. Math. **34**, 1365–1377 (1982)
5. M.N. Huxley, On the difference between consecutive primes. Invent. Math. **15**(2), 164–170 (1971)
6. M. Jutila, On the least Goldbach's number in an arithmetical progression with a prime difference. Ann. Univ. Turku; Ser. A. I **118**(5), 8 (1968)
7. A.A. Karatsuba, Sums of characters, and primitive roots, in finite fields. Dokl. AN SSSR **180**, 1287–1289 (1968) (Russian)
8. A.A. Karatsuba, Estimates of character sums. Math. USSR-Izv. **4**(1), 19–29 (1970)
9. A.A. Karatsuba, Sums of characters over prime numbers. Math. USSR-Izv. **4**(2), 303–326 (1970)
10. A.A. Karatsuba, Sums of characters with prime numbers. Dokl. AN SSSR **190**, 517–518 (1970) (Russian)
11. A.A. Karatsuba, The distribution of products of shifted prime numbers in arithmetic progressions. Dokl. AN SSSR **192**, 724–727 (1970) (Russian)
12. A.A. Karatsuba, Sums of characters with prime numbers in an arithmetic progression. Math. USSR-Izv. **5**(3), 485–501 (1971)
13. A.A. Karatsuba, Sums of characters in sequences of shifted prime numbers, with applications. Math. Notes **17**(1), 91–93 (1975)
14. A.A. Karatsuba, Some problems of contemporary analytic number theorem. Math. Notes **17**(2), 195–199 (1975)
15. A.A. Karatsuba, Sums of Legendre symbols of polynomials of second degree over prime numbers. Math. USSR-Izv. **12**(2), 299–308 (1978)
16. A.A. Karatsuba, Distribution of values of nonprincipal characters. Proc. Steklov Inst. Math. **142**, 165–174 (1979)
17. A.A. Karatsuba, *Principles of Analytic Number Theory*, 2nd edn. (Nauka, Moscow, 1983, in Russian), 240 pp
18. A.A. Karatsuba, Arithmetic problems in the theory of Dirichlet characters. Russ. Math. Surv. **63**(4), 641–690 (2008)
19. B. Kerr, On certain exponential and character sums. PhD Thesis, UNSW (2017)

20. Ju.V. Linnik, Some conditional theorems concerning binary problems with prime numbers. Izv. Akad. Nauk SSSR Ser. Mat. **16**, 503–520 (1952) (Russian)
21. Ju.V. Linnik, Recent works of I. M. Vinogradov. Trudy Mat. Inst. Steklov. **132**, 27–29 (1973); Proc. Steklov Inst. Math. **132**, 25–28 (1975)
22. K.K. Mardjhanashvili, An estimate for an arithmetic sum. Dokl. Akad. Nauk SSSR, **22**(7), 391–393 (1939) (Russian)
23. K. Prachar, Uber die Anwendung einer Methode von Linnik. Acta Arith. **29**, 367–376 (1976)
24. K. Prachar, Bemerkungen uber Primzahlen in kurzen Reihen [Remarks on primes in short sequences]. Acta Arith. **44**, 175–180 (1984)
25. Z.Kh. Rakhmonov, On the distribution of values of Dirichlet characters. Russ. Math. Surv. **41**(1), 237–238 (1986)
26. Z.Kh. Rakhmonov, Estimation of the sum of characters with primes. Dokl. Akad. Nauk Tadzhik. SSR **29**(1), 16–20 (1986) (Russian)
27. Z.Kh. Rakhmonov, The least Goldbach number in an arithmetic progression. Izv. Akad. Nauk Tadzhik. SSR Otdel. Fiz.-Mat. Khim. i Geol. Nauk **2**(100), 103–106 (1986) (Russian)
28. Z.Kh. Rakhmonov, On the distribution of the values of Dirichlet characters and their applications. Proc. Steklov Inst. Math. **207**(6), 263–272 (1995)
29. Z.Kh. Rakhmonov, Distribution of values of Dirichlet characters in the sequence of shifted primes. Dokl. Akad. Nauk Respubliki Tajikistan **56**(1), 5–9 (2013) (Russian)
30. Z.Kh. Rakhmonov, Distribution of values of Dirichlet characters in the sequence of shifted primes. Izv. Saratov Univ. (N.S.) Ser. Math. Mech. Inform. **13**(4(2)), 113–117 (2013) (Russian)
31. Z.Kh. Rakhmonov, Sums of characters over prime numbers. Chebyshevskii Sb. **15**(2), 73–100 (2014) (Russian)
32. Z.Kh. Rakhmonov, The sums of the values of nonprincipal characters from the sequence of shifted prime. Trudy Mat. Inst. Steklov **299**, 1–27 (2017) (Russian)
33. Z.Kh. Rakhmonov, On the estimation of the sum the values of Dirichlet character in a sequence of shifted primes. Dokl Akad Nauk Respubliki Tajikistan **60**(9), 378–382 (2017) (Russian)
34. Z.Kh. Rakhmonov, A theorem on the mean value of $\psi(x, \chi)$ and its applications. Russian Acad. Sci. Izv. Math. **43**(1), 49–64 (1994)
35. I.M. Vinogradov, Some general lemmas and their application to the estimation of trigonometrical sums. Rec. Math. [Mat. Sbornik] N.S. **3**(45), 435–471 (1938) (Russian)
36. I.M. Vinogradov, An improvement of the estimation of sums with primes. Izv. Akad. Nauk SSSR Ser. Mat. **7**(1), 17–34 (1943) (Russian)
37. I.M. Vinogradov, New approach to the estimation of a sum of values of $\chi(p + k)$. Izv. Akad. Nauk SSSR Ser. Mat. **16**(3), 197–210 (1952) (Russian)
38. I.M. Vinogradov, Improvement of an estimate for the sum of the values $\chi(p + k)$. Izv. Akad. Nauk SSSR Ser. Mat. **17**(4), 285–290 (1953) (Russian)
39. A.I. Vinogradov, On numbers with small prime divisors. Dokl. Akad. Nauk SSSR **109**(4), 683–686 (1956) (Russian)
40. I.M. Vinogradov, An estimate for a certain sum extended over the primes of an arithmetic progression. Izv. Akad. Nauk SSSR Ser. Mat. **30**(3), 481–496 (1966) (Russian)
41. W. Yuan, On Linnik's method concerning the Goldbach number. Sci. Sin. **20**, 16–30 (1977)

Printed in the United States
By Bookmasters